LAPWINGS, LOONS AND LOUSY JACKS

LAPWINGS, LOONS AND LOUSY JACKS

THE HOW AND WHY OF BIRD NAMES

RAY REEDMAN

Pelagic Publishing

Published by Pelagic Publishing
www.pelagicpublishing.com
PO Box 725, Exeter EX1 9QU, UK

Lapwings, Loons and Lousy Jacks: The How and Why of Bird Names

ISBN 978-1-78427-092-6 (Hbk)
ISBN 978-1-78427-093-3 (ePub)
ISBN 978-1-78427-094-0 (Mobi)
ISBN 978-1-78427-095-7 (PDF)

Copyright © 2016 Ray Reedman

The author asserts his moral right to be identified as the author of this work.

All rights reserved. No part of this document may be produced, stored in a retrieval system, or transmitted in any form or by any means, electronic, mechanical, photocopying, recording or otherwise without prior permission from the publisher. While every effort has been made in the preparation of this book to ensure the accuracy of the information presented, the information contained in this book is sold without warranty, either express or implied. Neither the author, nor Pelagic Publishing, its agents and distributors will be held liable for any damage or loss caused or alleged to be caused directly or indirectly by this book.

British Library Cataloguing in Publication Data
A catalogue record for this book is available from the British Library.

Cover image:
Lapwing Family by Martina Nacházelová (www.nachi.artstation.com).

CONTENTS

My father, Alf, taught me about the outdoors; my mother, Elsie, gave me my love of books. My wife, Mary, has been my companion and support for almost sixty years. My grandchildren, Robert, Jade and Amber, are the future. This is for them.

ACKNOWLEDGEMENTS

Many years ago, a colleague, Ann Cole, involved me in her work on 'eagle' place names (cited later). Another colleague, John Allen, gave me a copy of the *Oxford Book of Bird Names*. Both inadvertently sowed the seeds of this idea.

When I retired, the late Martin Sell was a generous mentor and friend who taught me a great deal and encouraged me to follow his lead in teaching the basics of ornithology to adults. Martin also introduced me to what is now the Berkshire Ornithological Club, where such as Colin Wilson, Ted Rogers and Bill Nicoll, among many others, shared their knowledge, skills and friendship. Ken and Sarah White later introduced us to the wonders of birding in Andalucia. It has been a fulfilling experience.

My brother-in-law, Reg Coombes, and his late wife, Jean, hosted us many times in Florida and Canada and always went out of their way to introduce us to local wildlife opportunities. My sister Marian and her husband, Henry Morrison, have always lived close to nature in Australia, so our visits to them have allowed me to savour some very special environments. As a result, so much of this book is about very happy memories.

In New South Wales, Clive and Enid Johnson were generous hosts: Clive introduced me to the name 'Lousy Jacks' as we explored their farm. In Nova Scotia, Bob and Wendy McDonald took us birding in the Wolfville area. More formal birding activities have introduced us to some wonderful professionals: Carol Probets in the Blue Mountains, Luke Paterson in the Northern Territory, Mike Crewe of Limosa, and Roodal and David Ramlal in Trinidad.

Carol was quick to respond to a couple of requests for help with Australian names, as too was Bob Flood on a pelagic matter. I much appreciate the help and feedback I had at an earlier phase of the work from Andy Swash and David Lindo in particular.

Ernest Leahy and the Birds of Berkshire Atlas Group have allowed me to reproduce the line drawing of a Common Scoter, which originally appeared in the 2013 Berkshire Atlas. Katrina van Grouw freely offered her wonderful Trumpet Manucode drawing, which first appeared in her masterpiece *The Unfeathered Bird*. The bird pictures are generally my own work, but there are a few exceptions: Adrian Brown supplied the photo of the Great Northern Diver/Common Loon; Mary took the photo of the Thekla Lark; and the late Ros Hardie photographed the Woodchat Shrike for us, as it was on her side of the minibus.

Clearly I am indebted to a whole library of references, whose authors for one reason or another have focused on the wonderful world of birds. These are listed at the end, but the works of Messrs Lockwood, Jobling, Choate and Macdonald have been particularly important to me. I much appreciate the permissions to cite from assorted works by the following: British Birds; the British Ornithologists' Union; the British Trust for Ornithology; Bloomsbury Publishing plc; Ebury Press; Encyclopaedia Britannica; the Gilbert White Museum and Thames & Hudson; the Harvard Common Press; HarperCollins; the International Committee for Zoological Nomenclature; Oxford University Press; Phaidon; Reed New Holland; and Tony Soper in person.

Adrian Brown and Gray Burfoot have long put up with my in-car ramblings as this book developed, and have both provided valuable feedback as reviewers of the text: in another life we all called it marking homework. Mary read the work through at an earlier stage. The final copy-editing by Hugh Brazier was a rigid and disciplined exercise, conducted with a real sense of common purpose and a great deal of good humour. As a result, rough edges were smoothed and errors corrected. Obviously I am indebted to Nigel Massen who willingly and enthusiastically took on the task of converting my draft into a finished volume. My thanks to him and to the Pelagic team.

And finally, none of this would be possible without the birds. This book is about much more than their labels, so please read on and I hope that you will discover that they have a much more profound meaning for me as symbols of wonder and beauty – and even of life itself.

PREFACE

I have always loved both words and birds. I have a background in language and literature, so when I arrived at a point in life where I could find more time for my hobby it was not difficult for me to be drawn to the language of, and the stories behind, bird names. I first became hooked on this topic when I started to visit North America in the early nineties: the handbook had different names for what appeared to be the same birds and I had to look more closely at the scientific names. The more I looked, the more the topic fascinated me. Then I realised that it was a shifting landscape and that it always had been. Visits to Australia and Trinidad brought me new experiences in other English-speaking cultures, where the names had their own local colour, and my fascination grew. The result is this book.

The narrative is not intended to be either a work of learned reference or, on the other hand, a bedside companion. Instead, I see it as a sort of bridge between the hobby of birdwatching and aspects of language which otherwise shut out the average birder. When I started to look closer at this subset of the world of birds, I found contradictions and confusions everywhere: linguistic, historical, cultural and national, as well as scientific. As a result I have assembled a compilation of comments and observations which, hopefully, both elucidate and entertain. At the least they should open a few doors into the complicated and confusing world of bird names, where those much better qualified as linguists, historians or scientists can be consulted: a reference list will be found at the end. What I want to share here is my enjoyment of the stories behind the words and, sometimes, how these link with my own experiences. For that last reason, I do not stray too far into realms which are foreign to me.

If you live long enough, it is increasingly likely that some 'facts' you learnt as a child are no longer the same. Find Southern Rhodesia and Tanganyika in a modern atlas if you can! It would be equally difficult to find a Pewit or a Green Plover in a modern field guide, yet all of those names were part of my childhood years (Pewit was a purely local variant of the more familiar Peewit, of course).

But time is only one of the confusions which face anyone entering the minefield of bird names. The accepted common names of birds can range from the obscure to the banal, from the romantic to the ridiculous. What is more, traditional names vary from region to region, with some variants still in use. The whole pattern is complicated by other English-speaking countries, which have their own traditional and formal differences to trap the travelling birder. Suffice it to say that there are myriad foreign tongues around the world, each with layers of quirks and traps, and more of us now travel to exotic corners in the pursuit of our interest.

As long ago as the eighteenth century, scientists decided to remove all the confusion by creating a universal pattern of names, which we know today as the Linnaean system. Since scholars communicated across borders in the classical languages, Latin and Greek formed the backbone of the system. However, what purports to be rational is sometimes surprisingly eccentric. Furthermore, times have changed and the common tongue of scholars now tends to be English. In any case, classical languages remain a mystery for the average birder, so we are still faced with confusion.

However, a major paradox puzzles the amateur, because the very scientists who required the consensus of the Linnaean system continue to muddy the waters. Almost every week we open a magazine or book to find that a familiar scientific bird name or relationship has been replaced or is being challenged. This particular channel turns out to be maelstrom, because speciation is constantly revisited using new criteria. It soon becomes obvious that even scientific 'fact' can be nebulous, and that the scientists have tenets and factions which contribute to the confusions. It is worth reading Nigel Collar's take on this (*British Birds*, March 2013). Mind you, dedicated 'listers' will find it depressing. In short, the minefield is more of the maritime sort, where the mines can drift around a bit. Between the first and the second editions of the *Collins Bird Guide*, the whole of the non-passerine section was confusingly realigned with the latest scientific findings, and 'new' species appeared. This was not the fault of the authors or editors, of course, but a necessary reaction to the changes wrought by the scientists in just one decade. I find myself caught in the very same dilemma, since taxonomic sequencing and the IOC list (the ever-changing authorative word of the International Ornithologists' Union) have been evolving even during the writing of this book. However, the book is about history, language and culture too, and the changing scene is part of its remit, rather than its problem. For pragmatic reasons I have anchored *The Names and the stories* to the shape of the second edition of the *Collins Bird Guide*, since that is familiar to most birdwatchers, while at the same time recommending that the IOC list be consulted for the most current news.

Apart from the scientific turbulence, we must accept that language is organic: it evolves with time, changed by the wear and tear of usage and by shifting fashions. Chaucer's *Parliament of Fowls* talks of *goos, cormeraunt, popinjay, throstil* and *feldefare*. Seven centuries later we can still recognise most of those easily, though *popinjay* has become obscure, while *throstle* lingers only in dialect and poetry. A close look at the twentieth century will show that some familiar names have had a bumpy ride in relatively recent history, among those Robin and Dunnock, while the twenty-first century has already seen the birth of an agreed IOC list of International English names, whose waves have yet to come fully ashore.

So do we take all of this as a serious obstruction to the enjoyment of our interest in birds? On the contrary, I see it to be the opposite, a rich mine of a different sort, one which bears seams of culture, history and linguistic curiosity just waiting to be exploited. The serious stuff is fascinating, but there's a lot of fun in it too.

My title hints at the complex range of issues to be discussed – common names, alternative usage, and the very local. Large numbers of scientific names are considered too. Consequently I have imagined the book as journey of exploration and discovery through an unfamiliar and confusing landscape, with fellow bird-lovers as my companions. For that reason it is a narrative which deals with the birds as we meet them.

The first two parts of the book contain an outline of linguistic and historical information which will help to set a framework of references. In the third part, I take a general look inside the scientific names. I see these three sections as the maps and compasses with which to travel. In the fourth part I take out my latest field guide and use its structure to explore the families of birds familiar in Europe, but I also relate the species discussed to those which I have encountered on other continents, notably in North America, Trinidad and Australia, where I have birded, but largely ignoring Asia and Africa, which I have not visited. And because those three areas contain so many new families, let alone species, I devote the fifth part to a further exploration of other names of those regions, which will round off the narrative.

I hope the trail will not prove too hot or too dusty. We will certainly find plenty of birds as we go. Bon voyage!

Ray Reedman

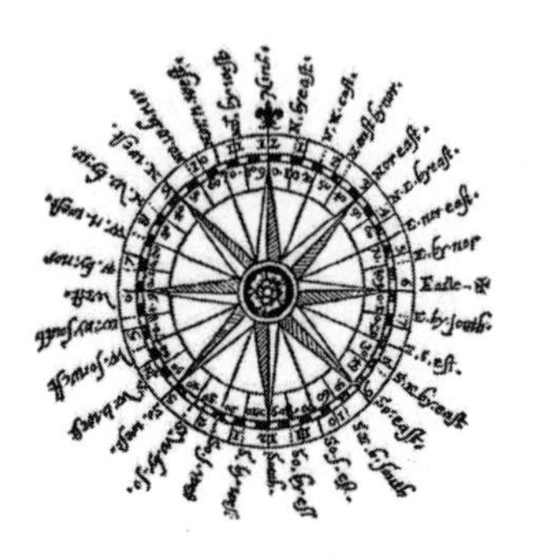

without system the field of Nature would be a pathless wilderness: but system should be subservient to, not the main point of, pursuit.

Gilbert White

HISTORICAL PERSPECTIVES

Language is a living entity which has its own roots and evolution. Society provides the mechanism of usage which drives those changes. This short section offers a brief and very general background to these matters.

THE SHAMBLES, YORK

ROOTS AND ROUTES

There will be frequent reference to the roots of words and to some of the routes taken by words towards their modern forms. It helps to realise that most modern European languages belong to an Indo-European group which includes Classical Greek and Latin, as well as those which we now refer to as Germanic and Romance (Italic) languages. Evolution in language happens a lot quicker than it does in birds, so words can change within a hundred years, and a lot more in a thousand, but the changes are often traceable. The diagram below shows the relationships between the main languages which contribute to this story.

In simple terms, Greek influenced Latin, because the later civilisation of Classical Rome looked to the Golden Age of culture and learning which had preceded theirs. The Romans created an empire which left a heavy legacy in the languages in Spain, Portugal, France and others, including modern Italy: these are the Romance languages, meaning that they were derived from the Roman tongue, spoken Latin.

Meanwhile, further north, the Germanic languages prevailed and included those of the Franks, the Goths, the Angles, the Saxons, the Jutes and the Norse among others. These were the peoples whose languages evolved into such as modern German, Dutch, Danish, Swedish and Norwegian. In Britain, and some other parts of Europe, an older Celtic civilisation left a legacy of Welsh, Gaelic, Breton, and others.

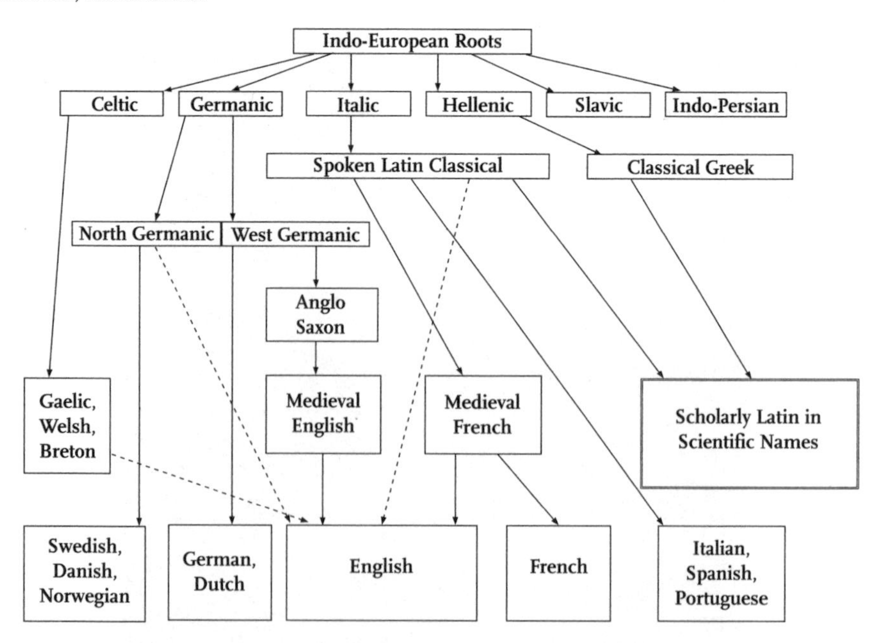

RELATIONSHIPS OF THE MAIN EUROPEAN LANGUAGES

The Roman occupation of Britain left almost no direct influence on our language, because any Latin remaining after the Romans left was erased by the subsequent influx into Britain of Angles, Saxons and Jutes. By the time the Vikings arrived, England was Anglo-Saxon, and spoke a Germanic form of language, to which Viking Norse merely added another and different layer, but mainly in the north and east of the country. The influence of trade and the Church had reintroduced small elements of Latin into Anglo-Saxon. The Norman Conquest of 1066 was part of a Viking quarrel (the word Norman had evolved from Norse Men), but the court of William the Conqueror had espoused the culture and language of France and brought to England a Latin-based tongue, Old French. For several hundred years, this was the language of the ruling classes, who were constantly reinforced by new blood from France, through marriage and inheritance. Slowly and of necessity this Romance language blended with the Germanic to give English an enriched form which was unique in Europe. A very simple illustration is that the living *cow* or *ox* (both Germanic in root) becomes *beef* (from French) when slaughtered.

This blend was reinforced later, when the Renaissance (the rebirth of Classical learning) came to Tudor Britain, bringing with it a more intense study of Latin and Greek. By the time Dr Johnson formulated his dictionary in the eighteenth century, Latin, which had always been the language of Christian scholars, was the universal language of lay-scholars in Europe and had been re-injected into the bloodstream of formal English. For the first time there was 'correct' English, a norm for the educated classes – and that was to become available to the entire population within a hundred years or so.

But we also have to remember that there was no single form of universal spoken English in Britain in the eighteenth century. Regional differences were often huge, and locally developed words and turns of phrase made for a series of sub-languages, or dialects. In these different forms there was much colour and variety, often now lost, largely because of universal education and the advent of mass media – radio, television and cinema in particular.

WIDENING HORIZONS

So what happened to the local names which existed in dialect? Some of them didn't go anywhere, of course: 'wheatear' and 'wagtail', for example, remain and have been elevated to name whole families. Other historical and local vocabulary travelled abroad to become established in other parts of the world: for example the Cornish 'murre' and the Scottish 'dovekie' both emigrated and survived in North America.

In Britain today we have lost many of the subtle differences which existed in a population of largely rural communities before the Industrial Revolution of the nineteenth century. Industry and technology slowly drew people away from the countryside and created a new mix of relatively mobile workforces. With greater urban integration came an increasing separation from nature and from its vocabulary. The dialect forms, which gave me the Pewit as a child, were fading into history. Words need to be used to survive, and even technology has its ephemeral vocabulary: 'cat's whisker' and 'tranny' are already in the museum of radio vocabulary, yet both had currency in my lifetime.

From the Renaissance onwards, Natural Scientists tried to make sense of the world of nature, and that included agreeing formal names in the English language. Even in this sphere of influence, names are adopted and then changed. My Green Plover existed for a while as the accepted name for what we now formally call a Lapwing, but the latter has now become the name of a whole family.

Universal education made society more literate and gave it access to books: the language of scientists and of amateur birders eventually met on the same ground. There was now a 'correct' form, which left the colourful variants of the past strewn by the paths of history. More recently, the Internet has given us access to information about pretty well any bird in the world, while air travel allows a new breed of 'world birders' to see them. We now need bird names in a worldwide dimension: Lapwing in my own records is now prefaced with Northern, Southern or Masked, so that I can distinguish the familiar home species from those seen in Trinidad and Australia.

But for the popular language all is not lost. We should be eternally grateful to Bill Oddie, who knows a good dialect word when he hears one: thanks to his frequent use of the term on television, the Shetland dialect name Bonxie is now widely used even by Sassenachs and Southerners, because it is so much more fun than Great Skua. Birders do in fact like colourful names – witness the formal retention of Dunnock after the attempt to impose Hedge Accentor a few years ago. In any case, the wheel is turning all the time: new dialect is being formed, as birders communicate in a language born of haste and technology: shorthand words such as 'barwit' and 'mipit' seem to prove that formality has only so much influence. Add to that the power of knowledge, and its vocabulary of 'larids' and 'hirundines', 'primary projection' and 'secondary coverts', and you find that the birding community has its own enclave, separated this time by culture, rather than by geography. A new birder may often feel like a foreigner in a strange land with its own language and dialect. But all is not lost: most birders are friendlier than they were a few years ago, and that may be because technology now tells us where to flock.

FRAMEWORKS

The activities of scholars led a transition from folk-names to formal names. This created a functional structural framework which today allows ornithologists and amateur birders alike to appreciate the complex world of birds. This section summarises the movements which created that framework.

EASTBRIDGE HOSPITAL OF ST THOMAS, CANTERBURY (PILGRIM LODGINGS)

THE MILESTONES

In discussing the content and history of bird names, certain reference points are important. There are several good histories of ornithology available, and fortunately it is not my mission to write another. In this case I need names and dates which are specific to the discussion of bird names. Much as I am a fan of Darwin, he had little influence on the naming of birds, so he is not included. Shakespeare was not an ornithologist, of course, but his use of language of the natural world provides some good evidence. A special word for Francis Willughby (1635–1672): he was the true naturalist in the partnership with Ray, but the latter outlived him to publish their joint work and his name is mercifully easier to type. Below are just some of the milestones along the road towards modern bird names. Some are historical events which bring about significant changes of culture and language. Some individual figures have significant influence on the shape of names, and some provide the evidence.

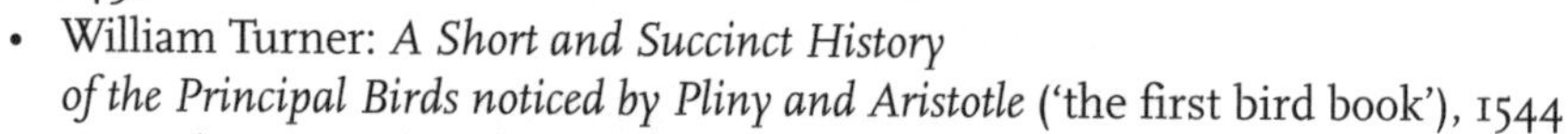

ARISTOTLE

- Aristotle, 384–322 BCE: the 'father' of natural history, whose work was the model until the eighteenth century
- Pliny (the Elder), 23–79 CE: *Naturalis Historia*
- Hesychius of Alexandria, 5th (?) century CE: Greek grammarian and lexicographer
- St Cuthbert, 635–687 CE: founder of the Farne Islands sanctuary
- Norman Conquest of England, 1066
- Albertus Magnus, ?– 1280: *De Falconibus*
- Geoffrey Chaucer: *The Parliament of Fowls* (many English bird names), 1382–1383
- Christopher Columbus landed on Trinidad, 1498
- William Turner: *A Short and Succinct History of the Principal Birds noticed by Pliny and Aristotle* ('the first bird book'), 1544
- Conrad Gesner: the bird volume of *Historia Animalium*, 1555
- Ulisse Aldrovandi (Aldrovandus): *Ornithology*, 1599–1603
- William Shakespeare, 1564–1616
- Jamestown, Virginia, founded, 1607
- The Pilgrim Fathers founded Plymouth, Massachusetts, 1620
- Christopher Merrett: *Pinax Rerum Naturalium Britannicarum*, 1666
- Walter Charleton: *Onomasticon Zoicon*, 1668
- Francis Willughby and John Ray: *Ornithology*, 1678
- Carl Linnaeus: 10th edition of *Systema Naturae* (binomial system), 1758
- Thomas Pennant: *British Zoology*, 1761–
- Botany Bay colony founded, 1788
- John Latham: *General Synopsis of Birds*, 1781–
- Gilbert White: *The Natural History of Selborne*, 1789
- Alexander Wilson: *American Ornithology*, 1808–1814
- Charles Lucien Bonaparte: major revision of *American Ornithology*, 1825–1833
- William Yarrell: *The History of British Birds*, 1843
- John Gould: *The Birds of Australia*, 1840–1848
- Alfred Newton: the foundation of the British Ornithologists' Union (BOU), 1858
- Elliott Coues/ Henry Seebohm: promotion of trinomials, late nineteenth century
- International Commission on Zoological Nomenclature, 1901
- Hartert, Ticehurst, Witherby & Jourdain: *A Hand-list of British Birds*, 1912
- Thomas Coward: *Birds of the British Isles*, 1920–1925 (BOU 1923 list)
- International Ornithological Committee (now the International Ornithological Union): the IOC list (definitive list of English bird names), 1990– (ongoing)

DEVELOPMENTS BEFORE LINNAEUS

English names represent our vernacular, which is of course paralleled in other languages and cultures. But the other half of the story is about the scientific names, which are Latinate, though not necessarily all Latin. Aristotle, long considered the father of natural history, was after all Greek. In truth, both languages were used widely by scholars throughout the post-Roman period. Before Linnaeus developed his *Systema Naturae* in the eighteenth century, the Babel of languages spoken throughout Europe was transcended by the culture of the abbeys and monasteries. The outposts of the Christian Church were oases of learning, which had evolved from the times of the Roman Empire, so for well over a thousand years all scholarship and much communication were in Latin, heavily underpinned by Classical Greek. The universities were founded in that tradition, which transmitted into the late-medieval Renaissance and onwards, through the Age of Reason and into the eighteenth-century Age of Enlightenment. Therefore Latin in particular was still the obvious and natural medium in which Linnaeus and his secular contemporaries wrote and communicated.

While the Renaissance was strictly the rebirth of Classical learning, the subsequent Age of Enlightenment was a period when the standards of the classical past were applied to modern reasoning, in science, political philosophy, language and the arts. It was the age of Thomas Paine, Voltaire, Dr Johnson, Isaac Newton, Benjamin Franklin and other such major figures, of which Carl Linnaeus was one.

Carl Linnaeus (1707–1778), was not the first to attempt classification: as David Attenborough put it, 'the first task of ornithology was to name birds'. This tradition in fact had its roots in the work of Aristotle, and many scholars had made attempts to build on this idea, though sometimes with rather odd logic, to say the least: the medieval Church had rules about eating only fish on Fridays, so it was convenient to classify some edible water birds as fish.

In the sixteenth century, scholars such as the Englishman Turner, the Swiss-born Gesner and the Italian Aldrovandus did much to promote the investigation and understanding of birds. In 1544 Turner produced what has been described as the first bird book. In 1555, Gesner's work included accounts of 217 birds. Between 1599 and 1603 Aldrovandus produced a monumental three-volume lifetime's work which named birds in four languages and classified them in a general way.

In 1666, Christopher Merrett published a highly incomplete and unreliable list, which nonetheless contained over 120 identifiable birds, including some binomial and even trinomial forms (see below). The list included names like *Aquila* and *Milvus*, which are still retained, but many which are not. *Pica* is there for the Magpie, but so is *Pica Marina* for the 'Sea Pye' (the Oystercatcher). In 1668 Charleton produced the *Onomasticon Zoicon*, which attempted to classify birds with reference to the work of his predecessors, and named the species in English, Latin and Greek.

This work was followed by a major contribution by Willughby and Ray, their *Ornithology*, published in 1678. In that work, based largely on Willughby's notes, Ray followed earlier examples by splitting water birds and land birds, but made further subdivisions, according to key features of appearance and habit. Much of the importance of this work lies in its influence on the development of the English names of birds.

LINNAEUS AND THE LINNAEAN SYSTEM

As stated above, there was much activity not only in Britain but elsewhere in Europe before Linnaeus produced a comprehensive framework which really worked. First published in 1734, and in eleven later editions, the *Systema Naturae* became the gold standard of classification and nomenclature. This is a monumental work which spans a huge part of the natural world: birds were just one part of it. Given that his name will forever be linked with the international system of naming all living things, it seems odd that in later life, upon his ennoblement in his native Sweden, Linnaeus changed his own name to Karl von Linné. In 1788, ten years after his death, the Linnean Society of London was founded, and to this day the society keeps Linnaeus's botanical, zoological and library collections.

CARL LINNAEUS

The system

One has to admire the logic which underpins the system, since it breaks down all living creatures into comprehensible chunks. Linnaeus developed a formula which really worked. Under that, the tree of relationships between each species can be traced. The parentage of the Arctic Loon/Diver works thus:

- Kingdom: Animalia (all animals, including mammals, birds, insects, etc.)
- Phylum: Chordata (vertebrate animals)
- Class: Aves (birds)
- Order: Gaviiformes: those resembling loons (divers)
- Family: Gaviidae (loons/divers)
- Genus: *Gavia* (loon /diver)
- Species: *Gavia arctica* (Black-throated Loon/Diver)

The final line for Linnaeus was the two-part (binomial) form, *Gavia arctica* (though, as will be discussed later, the uses of 'loon' and 'diver' are cultural issues between North America and Britain).

Trinomials

Linnaeus's was not the last word, however, since others continued to develop variants and alternatives. These were often controversial, with some views hotly disputed. In the late nineteenth century the concept of creating trinomials slowly emerged in America to eventually win wide favour: these provide names for subspecies (races). As a result, the Atlantic form of the Black-throated Diver is *Gavia arctica arctica* (the nominate race), while that of the north Pacific (Siberia and Alaska) is the subspecies *Gavia artica viridigularis*. Another example: the British race of the Yellow Wagtail is *Motacilla flava flavissima*, one of a number of subspecies of *M. flava*.

The International Code of Zoological Nomenclature (ICZN)

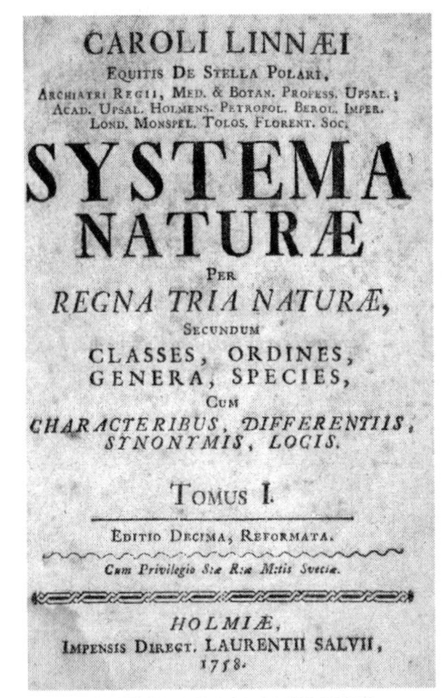
CAROLI LINNÆI
EQUITIS DE STELLA POLARI,
ARCHIATRI REGII, MED. & BOTAN. PROFESS. UPSAL.;
ACAD. UPSAL. HOLMENS. PETROPOL. BEROL. IMPER.
LOND. MONSPEL. TOLOS. FLORENT. SOC.
SYSTEMA
NATURÆ
PER
REGNA TRIA NATURÆ,
SECUNDUM
CLASSES, ORDINES,
GENERA, SPECIES,
CUM
CHARACTERIBUS, DIFFERENTIIS,
SYNONYMIS, LOCIS.
TOMUS I.
EDITIO DECIMA, REFORMATA.
Cum Privilegio S:æ R:æ M:tis Sveciæ.
HOLMIÆ,
IMPENSIS DIRECT. LAURENTII SALVII,
1758.

THE SYSTEMA NATURAE

The Linnaean system had survived more than a century and a half of debate and challenge when, in 1901, the scientists of the world met to establish the International Code of Zoological Nomenclature.

Draft rules were produced in 1905, but were soon sold out and readily became outdated. Evidently they were not adopted universally or immediately, since Hartert's *Hand-List of British Birds* in 1912 was largely provoked by the need 'to give each bird its correct scientific name in comformity with the Rules of the International Commission on Zoological Nomenclature'. Indeed Hartert's introduction was vehement in its condemnation of the inconsistency of many of his predecessors, revered or not: he named names and used phrases such as 'the evil of the want of uniformity'. In the quest for a new dawn, he exhorted his contemporaries to 'uphold the strict letter of the law'. In short, Hartert's list is a key moment in the conversion of British ornithologists to the rules of the draft code.

However, it took several decades more for the Commission to publish the first full edition of the Code proper, which eventually appeared in 1961. In its preface, J. Chester Bradley very neatly summarised the need for it thus:

> *Like all language, zoological nomenclature reflects the history of those who have produced it, and is the result of varying and conflicting practices. Some of our nomenclatural usage has been the result of ignorance, of vanity, obstinate insistence on following individual predilections, much, like that of language in general, of national customs, prides, and prejudices.*
>
> *Ordinary languages grow spontaneously in innumerable directions; but biological nomenclature has to be an exact tool that will convey a precise meaning for persons in all generations.*

With that latter thought in mind, the ICZN, as it is now known, was based firmly on the 1758 tenth edition of Linnaeus's *Systema Naturae*. In that way, the pre-eminence of the Linnaean system, complete with the development of trinomials, was finally underlined, and standardising rules for its use were established.

By that stage, the status of the Classical languages had already started to change, with scientists no longer using them as a major means of general communication: nineteenth-century European nationalism and empire-building on the one hand, and the growth of the United States on the other, had ensured that French, German and English vied with each other for supremacy on the international stage. That position has since changed to match the current patterns of world politics. Nonetheless, the Latin-based structure of the *Systema Naturae* survives to this day, complete with its indebtedness to Aristotle and Ancient Greece. As Tony Soper put it, 'the value of scientific names is that, being based on a dead language, Latin, they are not subject to the sort of change brought about by common usage which gives different meanings in different decades. The value of an ossified language is beyond price to the taxonomist who catalogues living creatures.' Nonetheless inaccuracies and slips do occur in the use of Latin or Greek, as we shall see later.

A while ago I was asked to introduce a multicultural class of seven-year-olds to the Kingfisher, and I used the opportunity to introduce the story behind the binomial, *Alcedo atthis*, which is recounted later. To put that name into context, I asked them to identify a dinosaur picture. The spontaneous shout of '*Tyrannosaurus rex*' convinced me that the Linnaean system is well embedded.

I will return to the details of the Linnean system a little later, but first let me look at the matter of how formal English names were shaped.

THE EVOLUTION OF FORMAL ENGLISH NAMES

Traditional names

As suggested earlier, vernacular forms of bird names are largely dependent on time and place. Indeed, Bircham points out that Willughby and Ray were sometimes confused in their travels by dialect names, and tended to duplicate, or to fail to make connections. Today it is much harder to realise that, in the past, even a small distance between populations could result in significant linguistic differences, and that certainly was reflected in the names used for wildlife. It is fading today, but I recall from my young days that three relatives, originating from three different parts of East Anglia, used the names 'twitch grass', 'couch grass' and 'spear grass' for the same garden nuisance, and that a fish known locally as 'nuss', was known up the coast as 'huss'. Natural scientists, like Linnaeus himself, often found themselves recording names given to them by mariners, crofters, market sellers and so on, and some of these were to find their way into the scientific names. Conversely, a great deal of language passes out of use with time, and many of the older names and forms become mere curios. It is tempting to dwell on more of these, but they may be found in the work of writers such as Lockwood.

Before the eighteenth century there was no such thing as a 'correct' bird name: regional variations were one thing, while the preferences, whims and creations of scholars made for a great variation in the vocabulary of scientific bird names, as will emerge later. Even spelling was arbitrary. In short, it was possible for naturalists

JOHN RAY

THOMAS PENNANT

to coin, adapt or borrow names according to their personal preferences. But the eighteenth century was to change all that radically, leading to the creation of a universal system of names which would be underwritten by all scientists by the start of the twentieth century. It would take a little longer to establish an authoritative voice on the English names of birds in Britain, and even longer to arrive at a system which would agree the forms of English names on the world stage.

The role of the ornithologists

After the sixteenth century, the formalisation of English vernacular names fell increasingly into the hands of the ornithologists. The names of such as Merrett, Ray, Pennant, Latham and Yarrell will appear time and again as their milestone-works steer names towards their modern forms: these ornithologists were certainly profoundly influential in the naming of birds.

Of these, Thomas Pennant is arguably the most important figure in this story alongside Linnaeus. Pennant may not be Britain's best-known naturalist, but his authority became stamped on the English names of many species which were previously in dispute or limbo. He began his series of volumes on *British Zoology* in 1761, and the formal confirmation of most English bird names coincides with his 1768 volume on birds. After that date, the majority (though not all) of his name forms were accepted as the norm.

From 1850 to 1857, the parson-naturalist Francis Morris produced *British Birds*, an affordable part-work publication, which, in a spirit of inclusiveness, listed alternative historical and popular names and also translated the scientific names. Such amateur efforts were not appreciated by the academic Alfred Newton, who founded the British Ornithologists' Union in 1858. Under him, this organisation became the self-appointed authority on the formalisation of British bird names, though the later influence of Hartert, Witherby, Ticehurst and Jourdain in the production of a *Hand-List of British Birds* in 1912 was significant.

In the preface to his *Dictionary of Birds*, 1893–1896, Alfred Newton referred to the 'contentious nature' of nomenclature. His contemporary, the art critic John Ruskin, clearly mocked such contention as pretentiousness when he wrote of a 'discussion of the reasons why none of the twelve names which naturalists have given to the birds are of any further use and why the present author has given it a thirteenth, which is to be universally, and to the end of time, accepted'. About three decades later, Thomas Coward commented that 'the quarrels of ornithologists are not yet settled; will they ever be?'

Matters certainly wavered and wobbled through the twentieth century, with no international consensus on bird names in the English language. For almost a century and a half such bodies as the BOU, the American Ornithologists' Union and the Royal Australasian Ornithologists Union struggled with issues of local preferences and institutional politics before attempting, within the framework of the International Ornithological Committee (now the International Ornithologists' Union), to agree formal international English names. A similar exercise by the French- and Spanish-speaking nations was rapidly concluded. However, work on the English names began in 1990, and took an astonishing fifteen years to complete. But it is now complete, is regularly updated, and can be consulted with ease online. The IOC mission statement for this project reads thus:

> *Our goal... is to facilitate worldwide communication in ornithology and conservation based on up-to-date taxonomy of world birds and recommended English names that follow explicit guidelines for spelling.*

With that outcome we have travelled down a path which has taken a hundred or so years longer than that taken for the scientific names. Three cheers!

But we are not really out of the woods yet, since Birdlife International still seems to be using its own version of bird names, which implies that we may well have to wait for harmonisation between the international bodies themselves. What is more, if we consult the current BOU list, we find that the organisation still maintains a parallel list of names for use at home, in which it differs from the

IOC list on such details as spelling and punctuation, as this small sample from their own table shows:

Scientific name 2016	**British (English) vernacular name 2016**	**IOC International English name 2016 ***
Nycticorax nycticorax	Night-heron	Black-crowned Night Heron
Pernis apivorus	Honey-buzzard	European Honey Buzzard
Regulus ignicapilla	Firecrest	Common Firecrest
Cyanistes caeruleus	Blue Tit	Eurasian Blue Tit
Panurus biarmicus	Bearded Tit	Bearded Reedling

* (Gill & Donsker 2016)

Clearly we need to be patient a bit longer. It is little wonder that the publishers of our bird guides still seem to be caught in a cleft stick over which is the name to use.

Over ninety years on from Thomas Coward's comment, it still seems that the maintenance of bird names is an ongoing process: indeed, the IOC *World Bird List* website speaks of 'the active industry of taxonomic revision', while its list of principles governing English-language names hints at areas of future revision. Maybe we are closer to a definitive set of names, but I am willing to wager that a hundred years from now they will probably look different again.

INSIDE THE SYSTEM

ST MARY'S CHURCH CLOCK, RYE, SUSSEX

For many amateurs, the world of scientific names is both exclusive and confusing, but it is rich in stories, quirks, mistakes and eccentricities, some of which will also appear later in the book. In this section I want to look at some of the general features which have been revealed during my hours of delving into the dictionary. So let us go and rummage around the inner sanctum.

THE MATRIX

To illustrate how the basic system works, it would be useful to look at two duck species.

In the case of the Marbled Duck, *Marmaronetta angustirostris*, the scientific name is composed of the generic name *Marmaronetta* (with capital M) and the specific name *angustirostris* (with a small letter). The first part of the name derives from Greek and can be broken down into words rooted in μαρμαρος (*marmaros*, marble) and νηττα (*nētta*, duck). The second part develops from the Latin *angustus* (narrow) and *rostrum* (bill).

The name of the Mallard, *Anas platyrhynchos*, comes from the Latin *anas* (duck), plus *platyrhynchos*, which derives from the Greek *platus* (broad) and *rhunkhos* (bill).

These two examples show just how readily the two languages are woven together to form the names: note that the words for 'duck' and 'bill' both appear in both languages. It has to be remembered that a lot of time had passed since Classical Greek and Latin had been spoken in the wider world, and that scholars had long used tricks which bent the Classical languages to their own needs as times had changed. From this, the term 'Medieval Latin' sometimes explains the origin of a word. Such tricks as used in the two examples above give quite a lot of elasticity to the system. The mixture of Greek and Latin words shows that it's Latinised, not Latin per se. In short, the language is a form of 'scientific Esperanto', which cobbles ideas together within a Latin structure. The system is utilitarian and is not intended to be linguistically elegant. It was born with the sort of flexibility which enables almost infinite modification to the elements of vocabulary around a firm and functional matrix. It is not anarchy, however: rules exist to govern the structuring of names, and for challenging the precedents of older forms, but as for the content and origin of words, almost anything goes.

Later on, I will be attempting to unearth some of the curios and stories which lurk in the convoluted thinking of those who created the scientific names which are encoded in sometimes obscure language. There are some wonderful tales – and some cause to feel that the authors of the names were, and still are, all too human.

CLASSIC DEVIATIONS

While Latin structure and some Greek vocabulary formed the basic framework of the system, it has not been possible to keep out other influences, many of which make for interesting stories. It is, after all, likely that scientists from non-European cultures would consider it reasonable to insert some local colour into these matters. For example, many of the proto-bird fossils found in China in recent years have a bit more than *sinensis* to liven up the new names. *Microraptor zhaoanis* and *Xiaotingia zhenghi* are two of many names which reflect this fact, with obvious local references.

Such deviation involves other languages too: *Falco cherrug* (Saker Falcon) derives its specific name from the Hindi word for the female falcon. The Australian *Falco berigora* (Brown Falcon) owes its specific name to an Aboriginal word. The name *Jacana* comes via Portuguese from a Central American Tupi word *yassana* – 'a noisy waterside bird'. (Having witnessed a spat between a Wattled Jacana and a Purple Gallinule, I can understand that.)

In short, it is possible to see almost any language crop up, and many will appear in later accounts.

THE NAMES BEHIND THE NAMES

One of the Chinese proto-birds (fossil ancestors) is known as *Confuciusornis* (Confucius bird). This acknowledgement of the great Chinese historical and spiritual figure is well in keeping with tradition. Personal names got into the *Systema Naturae* from the start, so once *Phalocrocorax aristotelis* (European Shag) had celebrated Aristotle it was difficult not to allow such dedications to proliferate.

Many of these were undoubtedly deserved by worthy scientists, explorers or patrons (though a fair bit of boot-licking happened too). That is fine if you have a name of an appropriate shape, but some names are just not suited to a Latinate form: *temminckii*, *verreauxii* and *hodgsoni*, for example, like Cinderella's sisters, just don't seem to fit the glass slipper. The worst example has to be *gulielmitertii* (a species of fig-parrot) which not only commemorates William III, King of the Netherlands, but squeezes in his number too! To my mind, the best of all such names is Cetti's Warbler, *Cettia cetti*, simply because Francisco Cetti (1726–1778) is recorded in stereo and, in case we don't cotton on, the bird shouts his name: 'cetti... cetti, cetti, cetti.' Lucky man.

One of the most intriguing stories was brought to my attention by the snipe which flew in front of our hire car in the mountains of Tasmania. The one true snipe in Australia is sometimes known as Australian Snipe or even Japanese Snipe, because it migrates between the two countries. However, its formal name is Latham's Snipe, *Gallinago hardwickii*. Why, I asked myself, was there a second person in the specific name, and who was Hardwicke? Jobling's *Dictionary of Scientific Bird Names* tells us that Charles Browne Hardwicke (1788–1851) was an English explorer and collector in Tasmania, so that would seem to fit the bill nicely. However, some consideration of the story of John Latham led me, intriguingly, to a link with the second man of that surname, Thomas Hardwicke (1756–1835). Latham (1740–1837), co-founder of the Linnean Society, and contemporary of Gilbert White, Joseph Banks and Thomas Pennant, has the reputation of being the 'grandfather' of Australian ornithology. It seems that he made the first descriptions of many species from the Cook expeditions, but failed to assign scientific names to many, including the snipe. In 1831 John Edward Gray (1800–1875), later keeper of zoology at the British Museum, finally named the snipe *Gallinago hardwickii*. At the time he was a colleague of Thomas Hardwicke, who was then working on *Illustrations of Indian Zoology* and whose real fame lies in introducing the Red Panda to the West. Whichever Hardwicke is concerned, it all seems a bit tough on Latham, but justice was done in other ways. In 1790, John White had named the Swift Parrot, *Lathamus discolor*, while the grandfather of Australian ornithology is also commemorated in the Glossy Black Cockatoo, *Calyptorhynchus lathami*, and the Australian Brushturkey, *Alectura lathami*.

But there is one personal name in the binomials which makes me shudder a bit. I have not been quite so keen on *isabellinus*, or the Anglicised form Isabelline, since discovering the story behind the word. It appears that the most likely explanation is that Queen Isabella of Spain (1451–1504) – better known as the joint sponsor, with King Ferdinand, of the voyages of Columbus – has been eternally linked to the fawn colour since vowing not to change her underwear until Spain was freed from the Moors. Somehow, I am no longer quite so anxious to see an Isabelline Shrike or an Isabelline Wheatear. The term is in evidence in French in 1595, and was used in 1600 to describe a gown in the wardrobe of Elizabeth I. It was widely used later in French to describe dun-coloured horses, so there is little surprise that its currency in bird names is down to its use by the French ornithologists Buffon, Levaillant and Vieillot.

SWIFT PARROT

Some personal names derive from mythology, so I feel that we need the Common Kingfisher, *Alcedo atthis*, on stage next to offset that last story. Atthis was a resplendent son of a Ganges river-nymph in Greek mythology. The juxtaposition of running water and splendid garb seems so apt for the bird, and proof that scientists can have a poetic soul (Hasselqvist in this case, and condoned by Linnaeus himself).

The Little Owl, *Athene noctua*, did a bit better than the Kingfisher, because it apparently latched onto the goddess herself, when a pair moved into the Temple of Athena to breed. In appropriate circumstances, this species will nest in crevices in rock, so a niche in a temple seems quite practical. This association elevated the owl to a status not often enjoyed in other cultures, where owls were, and sometimes still are, often seen as sinister.

THINGS IN THEIR PLACE

Some names may be misleading, and that of the Sandwich Tern, *Thalasseus sandvicensis*, is one. This is not a way of commemorating a Lord Sandwich, as the two-slice picnic does. For an explanation we must return to Latham, who named the species in 1797. The story goes that a collector from Kent sent skins of birds shot in his locality for Latham to describe and identify. Quite simply, Sandwich Tern is named after the town in Kent where the collector lived. The use of the name of a town is unusual – but there was a precedent, which contained something of a coincidence. A few years later, the same collector sent Latham an unusual-looking plover, which the latter christened Kentish Plover to acknowledge its source. However, the species had already been described and named by Linnaeus himself: *Charadrius alexandrinus*, after the city of Alexandria in Egypt.

The use of geographical names as such is otherwise quite common in bird names. Some, like *alexandrinus*, may also have echoes of the man behind the place name. Hudsonian Whimbrel, *Numenius hudsonicus*, for example, reflects the place, Hudson's Bay, rather than the man after whom the bay was named. St Francis too gets in on the act indirectly, for several Brazilian species have the name *franciscanus* – but only because they live in the area of the Rio São Francisco. There are many other such cases.

LITTLE OWL

Of course there are many broad geographical references which designate an area where a species was first found and named. One of the first introductions into Britain was the Common Pheasant, *Phasianus colchicus*, which the Romans brought with them, probably only as a domestically reared fowl, as the small number of archaeological records suggests. In fact, it is thought that the species was not widely distributed until the Normans had reintroduced it after 1066. The name of the bird is certainly much older than even the Romans, since it derives from the Greek word *phasianos*, after birds found by Jason's Argonauts on the banks of the River Phasis. As the river was in the country of Colchis, the second part of the name, *colchicus*, also reflects that ancient story. (The area concerned is today in the border zone between Georgia and Turkey on the shores of the Black Sea, and the Phasis is now the River Rioni.) Since several other races of Pheasant were eventually involved in the commercial business of Pheasant-rearing, only those without a neck-ring reflect the nominate race from Colchis.

Explorers like the Argonauts have always discovered species new to science, so given that Columbus's first landfall in the New World was in the Caribbean it is not too surprising to find that the Nearctic equivalent of the Common Buzzard, the Red-tailed Hawk, should be named *Buteo jamaicensis*. The species is widespread throughout North and Central America, yet it carries the name of a single island. By contrast, *Pluvialis dominica*, the name of American Golden Plover, has more to do with the hooded garb of Dominican friars than with the Caribbean island. The states of Carolina appear to feature in several names, such as *Anas carolinensis* (Green-winged Teal). The simple explanation is again historical and pre-dates the foundation of the two separate states, since the early British settlement there, founded in 1629 in the reign of King Charles I (*Carolus Rex*) was known as the Carolina Colonies, and therefore new species were named after it. In that way the first King Charles (in spite of an inglorious end) gets into the system in the same indirect way as Sir Henry Hudson and St Francis.

Two birds which I saw in Trinidad illustrate such specific local names for species which proved to be much more widespread: *Phalacrocorax brasiliensis* (Neotropical Cormorant) is one and *Leptodon cayanensis* (Grey-headed Kite) the other. Those I saw were a long way from Brazil or Cayenne (French Guiana) respectively.

One such name which intrigues me greatly is that of the Far Eastern Curlew, *Numenius madagascariensis*. I encountered one on a beach at Darwin, which underlines the fact that it has a far more easterly range than Madagascar. Was the first described specimen a vagrant when collected, or did the attribution result from confusion of specimens in transit? Such things did happen, as the story of the Kookaburra later illustrates. The species was in fact named as early as 1766 by Linnaeus himself, and had already been noted in 1760 by the French zoologist Mathurin Jacques Brisson (1732–1806), as *le corly de Madagascar*. The explanation is a simple one: there had been confusion between the place names Macassar and Madagascar. The place concerned was in the Celebes, now known as Sulawesi in Indonesia, and that makes far better sense, as it is on the species' migratory route to Australia, where many winter.

More sweeping generalisations such as Canada Goose, *Branta canadensis*, also occur, though some of its former subspecies have now been separated into full species, a reflection of the country's massive surface and of the diversity of evolution within that area. The word *americana* is surprisingly underused: I find it only rarely on my own lists from North America and Trinidad, in the names of Wood Stork, *Mycteria americana*; American Coot, *Fulica americana*; American Wigeon, *Anas americana*; Dickcissel, *Spiza americana;* and Tropical Parula, *Parula americana*.

Europe, of course, features very rarely, since the *Systema Naturae* was a European device. *Caprimulgus europaeus* distinguishes the European Nightjar from the others, and *Sitta europaea* distinguishes Eurasian Nuthatch. The odd thing is that the latter has several eastern races, so the trinomial comes into use: names such as *Sitta europaea asiatica* and *Sitta europaea sinensis* seem just a little eccentric. The term *asiaticus* tends to be reserved for species of the Indian subcontinent, which could possibly be explained by the fact that birds which are familiar in Europe are often Eurasian in their distribution.

If those eastern races of the Nuthatch are a mite puzzling, they fade into insignificance alongside *Cairina moschata* (Muscovy Duck). The association of a New World species with both the city of Cairo (in *Cairina*) and the city of Moscow (in both *moschata* and Muscovy) is bizarre in the extreme. It began in the late sixteenth century, when the duck was brought back to Europe for domestication. In simple terms, there was a very loose understanding of geography at the time and the association with those two exotic places was neither here nor there: both embodied the idea of 'exotic'. The once-used name, Cairo Duck, and its subsequent retention in *Cairina*, seems to be very definitely of that sort, while the leaning towards Muscovy may have had something to do with a trading company of that name. However, other theories involve two Native American peoples who kept the ducks domestically – the *Muiscas* of modern Colombia and the *Miskitos* of present-day Honduras and Nicaragua. Regardless of those more probable origins, the popular ideas eventually transferred into the scientific names of this species.

The word 'Turkey' has a similarly complicated history: this time the American species was attached to another exotic place, but via a secondary link with the guineafowls (themselves much more widely distributed than the single African country of the name). In the Middle Ages, the latter were imported into Europe via Turkey and were known in England as 'the Turkey bird'. The American species, introduced here some time around 1530–1550, was first thought of as a large guineafowl. The fact that it came from the west made no difference. When Linnaeus himself gave the species its scientific name, he complicated the situation even more by using the generic name *Meleagris* (which just happened to be the Greek for guineafowl). He then added Gesner's 1555 invention, *gallopavo*, which implied that the bird has characteristics of both domestic fowl (Latin: *gallus*) and peafowl (Latin: *pavo*). And just to make things worse, the word 'Turkey' later travelled to Australia to be attached to the Australian Brushturkey, which just happens to be a megapode.

While we are back in Australia, one of my great favourites is the New Holland Honeyeater, *Phylidonyris novaehollandiae*, a delightful jazzy bird which I first came to know in the Canberra region. A number of Australian birds have the word *novaehollandiae* in their scientific names. In the seventeenth century the Dutch were a major trading and colonial power in the islands which today contain Indonesia. Under their auspices, the first discovery of what is now the Australian mainland was made by the Dutch explorer, Abel Tasman, who named the area New Holland. Following the voyages of Captain Cook and the colonisation of New South Wales, the inhabitants of that area still knew the rest of the continent as New Holland, a habit which persisted well into the nineteenth century. As a consequence, new species were often named accordingly. The name Australia was coined long after use of the old form had become a habit. The use of *australis* to denote a uniquely Australian species is quite recent, as in the split of *Anthus australis* from Richard's Pipit.

Surprisingly, *africanus* appears to be used rather sparingly. For many of us, however, it is a word that we associate more with the Roman general, Scipio Africanus, the scourge of the Carthaginians. As far as I can see he had nothing whatsoever to do with the naming of birds. However, he did found the settlement of Italica near Seville in Spain, where there is a wonderful mosaic depicting the legend of Gerana and the Pygmies, which features later in the story of the cranes.

CARDINAL POINTS

Geographical references become truly generalised in some bird names with the use of a number of 'points of the compass' words: *septentrionalis, borealis* (northern); *meridionalis, australis* (southern); *occidentalis* (western); and *orientalis* or *sinensis* (eastern). Another of my Trinidad birds was *Buteogallus meridionalis* (Savanna Hawk), which implies that the bird belongs in South America. The Brown Pelican of the Americas is *Pelecanus occidentalis*, which makes sense because, from a European standpoint, it is the most westerly of the family. In Australia I encountered *Aythya australis*, the Hardhead, sometimes known as the White-eyed Duck, which has an Australasian distribution and might be considered a southern representative of the pochards and scaups.

Here in Britain, there is much debate among birders about the eastern race, *sinensis*, of *Phalacrocorax carbo* (Great Cormorant). *Sinensis* is a word in which the concepts of 'eastern' and 'Chinese' have been loosely muddled. In the name *Streptopelia orientalis* (Rufous Turtle Dove) we find another way of naming an eastern species. One of these graced Chipping Norton at the start of 2011 and was seen by so many twitchers that it needs no introduction from me –especially as I was not one of them.

And finally, on this topic, is there a longer and more cumbersome term than *australoabyssinicus*? It is a trinomial which locates a race of Bushveld Pipit, *Anthus caffer*, in southern Abyssinia (modern Ethiopia).

AND THE SCIENTISTS DIDN'T ALWAYS DO SO WELL...

First there is strange case of *Delichon*, as in *Delichon urbicum* (House Martin). In 1822, Heinrich Boie, an assistant to Temminck at Leiden, wished to underline the physiological differences between House Martin and Barn Swallow. To do so he coined an anagram of *Chelidon* (from the Greek *kelidōn*, swallow) which was, at that time, preferred for the swallow family. It is ironic that the more genuine form eventually lost out to the Latin *Hirundo*, and that the artificial creation *Delichon* still survives. All is not lost for *Chelidon*, though, as it still exists in *Petrochelidon*, the name of a rock-nesting swallow group, which includes the Cliff Swallow, *P. pyrrhonota*, which I saw nesting on various man-made 'cliffs': the buildings of Montreal and Halifax in Canada.

And this seems a good time to look at two owls: *Asio otus* and *Otus asio*. No, not a mistake, but a strange quirk of circumstances. One is our Long-eared Owl and the other the Eastern Screech Owl of North America: 'the Asio owl with ears' and 'the eared owl that resembles an Asio owl'. A certain lack of imagination there, one might argue. However, in recent years, the American genus has been renamed *Megascops*, so that story has been somewhat punctured. (*Megascops* is, of course, a large *scops*, which is a Greek word for 'a little eared owl'.)

Sometimes the story behind the names suggests that the learned men were a little uninspired, or tired, or that an idea had become a fixation. Take the Ruff, *Philomachus pugnax*, for example. Since the Greek *philomakhos* means 'pugnacious' or 'martial', there seems very little point in telling us that it is also *pugnax*, which means exactly the same in Latin. The Hoopoe gets the same treatment with *Upupa epops*, which is merely Latin for Hoopoe then Greek for Hoopoe. And this uninspiring habit is repeated with the Eurasian Curlew, *Numenius arquata*, both parts of which refer to the bow-shaped bill. Why the delicate and beautiful Grey Phalarope, *Phalaropus fulicarius*, finished up being likened to a Coot in both Greek and Latin seems a bit odd, until you take account of the *–pus* element (foot) of the Greek part, since it has 'coot-like' skin-flaps on its toes, which help it to swim. But why make the point twice?

FULICA ATRA AND *PHALAROPUS FULICARIUS*

Once again I have my own great favourite, this time taken from a lovely little bird which I have seen in Australia a few times, the Spotted Pardalote, *Pardalotus punctatus*. Suffice it to say that one of Shakespeare's 'seven ages of man' sees him 'bearded like a pard' (leopard). This derives from the Greek *pardalōtos* (spotted). Now, *punctatus* is Latin for – guess what? You got it in one. So my Australian bird translates as 'spotted' in all four parts of its name and in three languages. It doesn't come sillier than that. I spotted my first one in the Botanic Gardens in Canberra, by the way.

SPOTTED PARDALOTE

WARTS AND ALL

So far I have considered just a few of the characteristics and curiosities of the scientific names. Most are are logical and helpful, but some are bizarre, some are puzzling and others contain unhelpful errors. Many more such curios will be revealed in the remainder of the narrative. Linnaeus himself got a few things wrong, but sometimes that was because he was working on the basis of information that was erroneous when he inherited it from others. One of the basic rules in the reconsideration of scientific names was expressed on an American University website as the 'oldest fool wins', which is an irreverent way of saying that, even if an old name has misconceptions or errors, it has a historical precedence. In that way the Green Sandpiper retains its erroneous yellow legs (see later notes on waders) and the Far-eastern Curlew is still wrongly attached to Madagascar. There are a great many such examples which are retained to save creating even greater confusion.

In summary, there is plenty of evidence that scientific bird names contain a range of curious features which make them worth a closer look. From a wholly rational point of view the names contain many flaws and a fair bit of nonsense. But there is a limit to rationality, and the very fact that so many passions are aroused in the naming of birds suggests that scientists work with more than their heads. In the end, the system matters, but, for the purposes of classification, its content is not less efficient for containing quirks, eccentricities, colour and even downright error. For me, clearly, that is part of the charm of the system. In any case the system is secondary to the natural world which it classifies. As Gilbert White put it:

> *without system the field of Nature would be a pathless wilderness: but system should be subservient to, not the main point of, pursuit.*
> *(Letter XL to Barrington)*

White was clearly referring to the practical daily pursuit of studying, understanding and appreciating the environment around Selborne. I agree with him wholeheartedly and would go as far as to say that he has inspired my love of nature rather more that Linnaeus ever has, because his acuity of observation, his questioning and his sense of the wonder of the natural world have long been an influence. However, I do appreciate the framework which 'the system' provides, and from the point of view of this narrative it is an important part of a much wider story. Elements of it will therefore be woven into the subsequent pages, where I deal with the specific aspects of British bird names, and some of the names of those I have met on other continents.

THE NAMES AND THE STORIES

THE ROUND TABLE, GREAT HALL, WINCHESTER (CHRISTOPHE FINOT)

In this part of the book I look at birds by family and deal with common names and scientific names in tandem. The objective is to create a narrative interpretation of the dictionaries and the histories.

As for the organisation of that narrative, I find myself challenged to devise a structure which works: I don't want to create Gilbert White's 'pathless wilderness', but at the same time it is obvious that the whole idea grew out of confusion and that I am constantly trying to beat a path through linguistic and scientific nettles and brambles. Not least of these tangles is the question of the changing scene of taxonomic sequencing, so for the sake of simplicity I have used my second edition of the *Collins Bird Guide* as the main reference, because it is familiar to so many people. However, like any good birder I will wander down a few side-paths as I go.

My mind wanders a bit too, because memories of important moments will populate the stories. From time to time I indulge in a phenomenon which I have dubbed 'typographical evolution': all it takes is a clumsy keyboarding technique, and new birds appear. In a curious way, it is a valid illustration of what sometimes happened in the evolution of names in reality. However, rather than waste a good accident, I tend to invent the bird to go with it. Bear with me.

One of my returning themes will be just how readily Britain can be divided from North America by the same language, so here is a bird-related scenario which represents an amalgam of the ways in which we can differ without trying. Scene: Preparations for a day's birding on both sides of the pond:

> *I want your gear/stuff in the trunk/boot by 8 a.m. After the gas/petrol station we hit the interstate/motorway and drive to the preserve/reserve to be in the blinds/hides by 9 a.m. There will be plenty of peeps/sandpipers and other shorebirds/waders. We should get Black-bellied/Grey Plovers, possibly loons/divers, maybe some jaegers/skuas, and, with luck, some murres/guillemots offshore...*

So when I compare our different uses of the English Language, I try not to judge. We have to acknowledge that what prevails is cultural on both sides of the Atlantic, and that the macropolitics of business between the AOU and BOU, for example, will be complicated by micropolitics within each institution. In short, getting agreement in harmony and unison on issues of the English-language names for birds is not always that easy. We can all get a bit parochial and possessive, but I find American and Canadian birders great people to be with, because we all share a passion. It should be a topic for amusement rather than for national pride.

But Americans really do say things somewhat differently.

YOU MAY NOT HAVE A LEG TO STAND ON IN COURT

WILD ABOUT WILDFOWL: SWANS, GEESE AND DUCKS

Is there a better term than 'wildfowl'? The IOC list currently uses the term 'waterfowl', but I have always taken that to mean other water birds, like coots and grebes. The term wildfowl is an old habit, because I have to confess to having been a wildfowler of sorts. In the postwar years we lived on a farm for a while and, during rationing, my father used a shotgun to supply our table. My participation grew out of that, but was over quickly enough when I realised that I was enjoying the watching more than the shooting. In that, at least, I followed Sir Peter Scott's example. Suffice it to say that I am still unhappy that the wildfowlers' term should be the simplest name to cover the family. The formal term, Anatidae, is at least neutral, but it hardly seems practical, so wildfowl it has to be, unless we prefer to use the three words: swans, geese and ducks. But those three terms are not so innocently simple: even in Britain, there are several categories of duck: 'dabblers', 'divers', 'sawbills', 'stifftails', 'sea-ducks', some of which terms overlap. The shelducks are not true ducks: they sit in a place somewhere between ducks and geese. And within the geese, some are 'grey geese' and some are 'black geese', and some are a bit iffy. So what do the swans have to offer?

Swan song

The three British swan species offer rather more of a tangle than might at first appear...

Historically, the term 'wild swan' was used to describe the migratory species, and 'tame swan' referred to the sedentary species. When Pennant learned in 1785 that the 'tame swan' bred in Russia, he clearly felt that it was not so tame after all. To reflect this more accurately he created the names Whistling Swan for the former and Mute Swan for the latter. The fact that the former does not whistle led to the preference by others for an older name, Hooper, which had been known since 1566, and that was the name used by Yarrell in 1843. Later in the century the preferred spelling became Whooper. In 1883, the BOU standardised it as Whooper Swan.

The term 'tame swan' was rooted largely in the fact that monasteries, such as Abbotsbury in Dorset, fed swans to keep them close by as a food source, a privilege granted by the Crown, which formally claimed ownership of all swans in the twelfth century. Any local park will well illustrate just how easily the species is tempted to food (though, these days, the same can be said of many wild swans at a WWT centre). Yalden and Albarella cite archaeological evidence which proves that the Mute Swan is indeed a native species, not an introduction as had sometimes been thought. The Mute Swan is not entirely mute, of course, but Pennant did prevail on that one. As for its scientific name, *Cygnus olor* – the generic part derives from the Greek *kuknos* (swan) while *olor* is the Latin for... swan. The Whooper is simply *Cygnus cygnus*.

In 1830, Yarrell had identified the smaller 'wild swan' as a separate species and dubbed it Bewick's Swan in honour of the recently deceased engraver Thomas Bewick. That name was subsequently generally accepted. However, Bewick's is now considered to be a race (*bewickii*) of the Tundra Swan, *Cygnus columbianus*, with the North American (nominate) race distinguished by the popular name

Whistling Swan. In that way Pennant's old name travelled with the settlers and lives on. The word *columbianus* derives from the Columbia River, Oregon.

BEWICK'S SWANS

From *cygnus* came the French diminutive form 'cygnet' for the young swan. The male is the 'cob', a word which links with an alternative use as 'big man, leader', while the female is the 'pen', which *OED* suggests is of unknown origin. However, I have always had the facetious mental image of a monk sneaking up on the less-aggressive female swan to pinch his next quill. There certainly is likely to be a link with the Latin *penna* (feather), and I have seen it suggested (with no substantiation) that there may be a link with the fact that the pen keeps her wings to her sides (compare 'pinioned'), while the cob raises his in display.

The word 'swan' itself derives from an old Germanic form, which Lockwood traces back even further to link it with the concept of 'sound', which he suggests may derive from the unusual sound of the Mute Swan's stiff-pinioned flight – a flight of fancy, one might say. But there are Ancient Greek legends which do link the Mute Swan with song: mute in life, but singing beautifully just before death. The legend was refuted vigorously by the Roman naturalist Pliny, but there is no killing a great poetic image: it persisted to be used by Chaucer, Shakespeare and Tennyson among others, and the phrase 'swan song' enters the English language to become the metaphor for a last positive action. Coleridge's poem 'On a Volunteer Singer' has the best take on the story, since he explains why I don't watch talent shows:

Swans sing before they die – 'twere no bad thing
Should certain persons die before they sing!

There are remarkably few swan species in the world, and the three species which are discussed above represent almost half of those. In recent years, sufficient numbers of the Australian Black Swan, *Cygnus atratus*, have escaped from collections for there to be a small breeding population in Britain: in fact there was breeding evidence from more than 100 tetrads during the 2007–2011 BTO Atlas surveys. This raises the question of when 'escaped' birds become 'feral', and this species must surely be on the cusp of that decision. Its specific name, *atratus*, simply echoes the meaning of the English name. While I have seen such birds regularly in recent years, there is nothing quite like seeing them in a genuine Australian setting.

Goosy, Goosy, Gander...

'Goose' and 'gannet'... Why link these two words? The simple answer is that they both seem to have originated from a common Germanic root *gans*, which gives rise to both goose and gander in Medieval English, and later leads on to gannet, which I will look at again later. Lockwood goes even further back by suggesting that an older Indo-European root *ghans* was also behind the evolution via Greek to Latin and the resultant *Anser*, which we use today as the genus for many goose species. Gosling is structurally related to words like Sanderling and Starling (see later), with a diminutive suffix.

Geese have long been domesticated, and the species forming the root-stock of the European farmyard goose was the Greylag. There was a long tradition of setting out grain to lure birds into net traps, into decoys, or to within bow-range. It doesn't take much imagination to see the geese lured into domesticity via these situations. Selective breeding from mutations such as white birds would have followed, particularly as that mutation was prized by the Romans. Today a high percentage of the Greylag population in Britain originates from escaped or released stock, with some birds still showing a touch of the farmyard, in their white or patchy plumage. Truly wild Greylags still migrate from Iceland into Scotland in winter, of course.

The word Greylag itself has an interesting story: Lockwood feels that the element *–lag* has very ancient roots and that its original sense was 'goose', Greylag being a version of 'grey goose', which makes our modern usage, Greylag Goose, a tautology. A 1911 *Encyclopaedia Britannica* entry on the topic offers the following:

> *Skeat suggested (Ibis, 1870, p. 301) that it signified late, last, or slow, as in laggard, a loiterer (etc...).Thus the grey lag goose is the grey goose which in England when the name was given was not migratory but lagged behind the other wild species at the season when they betook themselves to their northern breeding quarters.*

Coward (1926) quotes Harting as linking the word to Old English *lea* (field) and Middle English *lac* (lake), a theory which he seems to acknowledge in his own version of the name as Grey Lag-goose. Lockwood notes that, in medieval times, goose flocks were summoned by the call 'lag-lag-lag' and such a usage is attributed by the encyclopaedia article to more modern Lincolnshire goose-herders. (The big question is whether the geese, when called, bothered to 'anser'.)

The name of the Bean Goose, *Anser fabalis*, is not as simple as it might seem. Lockwood tends to favour the idea that it fed among crops of field beans. It was also known as Corn Goose for similar reasons. Pennant had suggested in 1768 that the dark nail of the bill resembled a horse bean. In either case, the answer lies in a literal use of the word. This is reinforced by *fabalis*, the Latin for broad bean being *faba*. Today there are split beans in the offing, because birders now increasingly distinguish Taiga Bean Goose and Tundra Bean Goose, which represent two separate populations from different habitats in different geographical areas, and with visibly different builds and features. Though not yet recognised by the BOU, the split made by the AOU has already been acknowledged in the IOC list, where the Tundra Bean Goose is now flagged as *Anser serrirostris*, meaning 'saw-billed'.

Also distinguished in the field (though less easily) are two races of the Greater White-fronted Goose, *Anser albifrons* (whose scientific name neatly supports the vernacular form, with *frons* and 'front' indicating the bird's forehead). The two races seen in Britain are known as Greenland and European, though most of us ignore the 'Greater' element of the full international name, since its smaller relative, Lesser White-fronted Goose, *Anser erythropus*, is so rare. The word *erythropus* (literally 'red-legged') is a pretty feeble name, as Hartert pointed out when he preferred *finmarchicus* (from Finnmark, Norway), which did at least indicate the area of its breeding range. It might have been better to feature the orange-yellow eye-ring, which helps to separate this species from its larger cousin.

The Pink-footed Goose, *Anser brachyrhynchus*, has a vernacular name which distinguishes its leg-colour from that of the other grey geese, though in the field this is unreliable. Far better is the hint conveyed in *brachyrhynchus* (short-billed), as the neater head is relatively easy to distinguish at distance.

All of the above species are of the genus *Anser*, and are collectively known as 'grey geese'. The genus *Branta* has its own collective form: since Barnacle Geese and Brent Geese are broadly similar, they were often lumped together as 'black geese', a term which was still used by Essex wildfowlers in my youth. This almost certainly accounts for the fact that Brent Goose finished up, erroneously, with the barnacles in the name *Branta bernicla*. Both Brent and the generic name *Branta*, and of course the international name Brant, are derived from the Norse *brandgas*, meaning 'burnt goose', a reference to the black coloration. The word always reminds me of those old cinema Westerns, where the branding irons burn the ranch brand into the hide of cattle. Used to mark trade goods, the technique was the origin of 'brand names' in commerce.

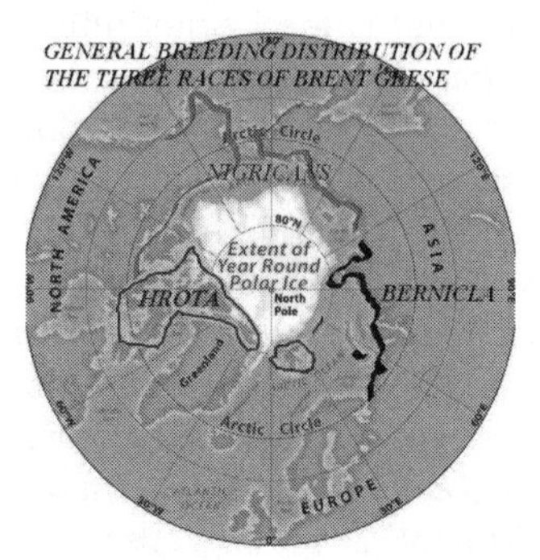

BREEDING DISTRIBUTION OF BREANT GEESE

Brent Geese have a circumpolar breeding range, which has resulted in the evolution of three distinct subspecies. The most numerous in Britain, though not in Ireland, is the Dark-bellied nominate race, which breeds in western Siberia.

The Pale-bellied race, *B. b. hrota*, also winters in Britain, and especially in Ireland, in good numbers, but breeds from eastern Canada, through Greenland and Svalbard to Franz Josef Land. The Icelandic language is responsible for the echoic form *hrota*, which derives from *hrot goes*, meaning 'snoring goose', a reference to the distinct call of the species.

Black Brant is the name preferred in North America for the race *B. b. nigricans*, which breeds in eastern Siberia and western North America. These occur in Britain in very small numbers, but annually, as individuals are 'carried' by flocks of the two regular races, possibly from both the east and the west. The adjective *nigricans* simply means 'blackish'.

'Carrying' occurs when disoriented individuals attach themselves to a larger migrating flock of another similar species or subspecies and are carried in an abnormal direction. One such, which is sometimes carried by Dark-bellied Brent Geese, is the Red-breasted Goose, *Branta ruficollis*, whose name means 'red-necked', which is also true, since the brick-red colour extends into both neck and breast.

RED-BREASTED GOOSE WITH DARK-BELLIED BRENT GEESE

As for the Barnacle Goose, *Branta leucopsis*, the story of its vernacular name is much more in the realms of Celtic fantasy, since there was a need to explain the seasonal appearance and disappearance of the birds. A form of shellfish was seen as the source of the returning geese, with the birds apparently hatching from the goose barnacle. The legend was first recorded by Giraldus around 1185 and persisted until the seventeenth century. Such folklore was common enough, and even Gilbert White was still unsure, in the eighteenth century, about the belief that swallows hibernated in mud. From a twenty-first-century point of view, this all seems a little incredible, but we have to remember that belief, fantasy and legend often prevailed in a different way from today. Mind you, there may have been more to the barnacle idea than meets the eye, since it may well have been linked to the Catholic prohibition of all meat on a Friday: 'fish' was allowed. The species name, *leucopsis*, derives from the Greek *leukos* (white) and *opsis* (faced).

BARNACLE GEESE

The introduced Canada Goose, *Branta canadensis,* is probably the most familiar species in Britain, though there are many who would wish it otherwise, since they are bulky, noisy and often too numerous. These are of the nominate race of what is sometimes referred to as a 'complex'. In 2004 the smaller representatives were split to form a separate species, the Cackling Goose, *Branta hutchinsii,* which is named for its distinctive call, and after Thomas Hutchins (1730–1790) of the Hudson's Bay Company. On average these are about the size of a Barnacle Goose and therefore well under half the size of the Canada Goose as we know it. For that reason they are more likely to be claimed as truly wild vagrants if and when they appear among other species in Britain. Most often noted are the forms known as Taverner's Goose and Richardson's Goose.

In North America there are also three species of the genus *Chen* – from the Greek *khēn* (goose) – which is also accepted by the IOC. The Snow Goose, *Chen caerulescens,* appears on the British list as a vagrant, but the other two are considered to be escapes when they occur. Both the BOU and the *Collins Bird Guide* still prefer *Anser* to *Chen.* The specific name *caerulescens* refers to the 'blue' morph which occurs frequently in the more numerous subspecies known as Lesser Snow Goose. Smaller still is Ross's Goose, *Chen rossii,* which name was dedicated to the Hudson's Bay Company's Bernard Ross as a gesture of gratitude by the American ornithologist John Cassin (1813–1869). Ross had organised the safe transportation of collector Robert Kennicott's specimens after his tragic death, aged just 30, during an Alaskan expedition in 1866. That must surely be one of the most unusual reasons for the attachment of a person's name to a species. The Emperor Goose, *Chen canagica,* takes its specific name from Kanaga Island in the Aleutians. Choate offers no help with the origin of 'Emperor'. The equivalent name is widely used in other languages, and I suspect that it was allocated for the bird's generally regal appearance, with more than a hint of the drape of the robe of a Mikado in the contrast of the clean white head and neck with the dark body.

Like the Canada Goose, the Egyptian Goose was introduced into Britain as a collection bird, but soon escaped and became established in the wild. It represents a link between true geese and the shelducks. Its colour is subtly conveyed in its generic name: *Alopochen* relates closely to the Greek name *khēnalōpēx*, which it apparently shared with the Ruddy Shelduck. Both words derive from *alōpēx* (fox) and *khēn* (goose) to describe the rufous colour of both birds. In order to create the scientific name of the goose, the scientists simply reversed the two elements. Tacked on to *Alopochen* is the obvious geographical word *aegyptiaca*.

Australia has two very distinctive geese, the first being relatively closely linked to those of the northern hemisphere. Cape Barren Geese belong to the unique genus, *Cereopsis*, which means 'wax-faced' and refers to the very striking cere, which sits on the bird's bill like a blob of lime-green wax. *C. novaehollandiae* is as Australian as they come, since it inhabits a limited range of the southern coast of mainland Australia and Tasmania. The island which gives them their name is between Tasmania and the mainland, but it was in the south of Tasmania that we used the car as a hide to approach a grazing flock of these beautiful geese.

The second species is not really a goose at all, since the Magpie Goose, *Anseranas semipalmata*, belongs in a family of its very own, the Anseranatidae. It is in fact a primitive sort of living fossil, a species which diverted from the ancestor species before the evolution of the true ducks and geese: indeed the scientists sit on the fence with a name which includes both *anser* (goose) and *anas* (duck), while *semipalmata* points out that the species is not even fully webbed. It has other curious features too, not least the fact that one male and two females will form a family unit to raise young together. We met them in huge numbers in Kakadu, since it is a truly tropical species.

MAGPIE GOOSE

Before I leave the geese behind I have to confess that the Brent Goose once evolved in my notes into Bent Goose. This name struck me as appropriate for a new race which will evolve among the offshore wind farms. And that also reminds me of another typo goose which many twitchers may know, in spirit at least: the Been Goose is the one that has just left when you arrive to see it.

Neither shield nor shell

The shelducks are strictly neither goose nor duck, but occupy a slot somewhere in between the two, with the Egyptian Goose somewhere close by. They are all strikingly colourful species in which the sexes are broadly similar, though more easily distinguished than is the case with geese.

The Common Shelduck, *Tadorna tadorna*, is certainly one of the most striking of wildfowl. Lockwood explains that the name Shelduck is more correctly Sheld Duck, since the bold colour patterns of the bird are at the root of a word meaning 'variegated'. He does, however, point out that the oldest evidence of the name occurs in the personal name Sheldrac in 1195, with that form first recorded as a bird name, *Sheledrake*, in about 1325. 'Shell Duck' first appeared in 1707. In tracing the root of an alternative folk-name, *Bergander*, he suggests that the root of *ber–* might be as in an Old Norse word meaning 'berry', therefore the berry-red horn on the beak of the male, while *–gander* is the male form, as in geese. In that he implies that there was a bias towards the male form in *Sheldrake*, which was almost certainly the younger of the two words. What is certain is that there is no justifiable link in the name to either 'shell', which might be reasonable for an estuarine species, or to 'shield', which has sometimes been supposed with reference to the knob on the bill.

The generic name, *Tadorna*, was based on a French name for Shelduck. That in turn has an Italian origin. Five other species of shelduck exist in Eurasia and Australasia. Ruddy Shelduck, *T. ferruginea* (rusty), appears as an escape and possible vagrant in Britain, while Australian Shelduck, *T. tadornoides* (resembling *tadorna*) is to southern Australia what the Raja Shelduck is to the north of the continent. There is no representative in the Americas. The beautiful *Tadorna radjah* has a name which originates as a Moluccan word *Radja* for the species. In Australia it also goes by the popular name of Burdekin Duck, after the name of a river in Queensland. We saw them in Queensland and in much larger numbers in Kakadu, where the pristine white of the birds was often stained by the mud in which they foraged.

Ducks are a-dabbling, up-tails all...

The word 'duck' derives from an Old English word meaning 'to dive', though some are exclusively surface-feeders and are known as 'dabblers'. Most British surface-feeding ducks belong to the genus *Anas*, the Latin for duck.

The name 'wild duck' was long used for the most common species, though the word Mallard existed alongside it for centuries. Its roots were in the Norman French, *maular*, where it may have originated from *masle*, meaning 'male' (*OED*). It acquired the suffix *–ard* (which is found in Poch*ard* and Buzz*ard* among others) and tended to distinguish the drake in particular. The present form was preferred by Pennant in 1768, though both Yarrell, 1843, and Newton, 1896, insisted on the alternative Wild Duck. The matter was settled when Hartert *et al.* preferred Mallard, but the entry in Coward's 1926 version of the BOU list still retained the entry 'Mallard or Wild Duck'. Old habits die hard, but I do not recall the term Wild Duck being used in that specific way during my lifetime, even by an older generation of wildfowlers.

The scientific name, *Anas platyrhynchos*, means 'broad-billed duck'. Coward notes that 'the time-honoured name *A. boschas* is rejected because Linnaeus named the female *platyrhynchos* two pages before he named the male!' The alternative form was based on the Greek for wild duck, but the current form had been in use since 1758, presumably because it was the earlier entry in the *Systema Naturae*. The 'rejection' was pronounced in a judgement by the BOU in 1915, but the real importance of the matter is that the two sexes had been treated separately – an issue rooted in the word Mallard itself.

The names Gadwall, Teal and Garganey all seem to have originated in echoic forms. Such an origin for Gadwall is more easily understood in an earlier form, Gaddell, which was noted by Merrett in 1667 (cf. to gabble). The scientific name *Anas strepera* (noisy) supports that concept. Ray and Pennant both adopted the current spelling, which became standard.

GADWALL

'Teal' first appeared in 1314 as *teles*, with roots probably in an earlier Germanic form based on the whistle of the male. The name was used by Merrett and by Ray, and formalised by Pennant in 1768 in its present spelling. The scientific version *Anas crecca* originates in the Swedish name for the species, *kricka*. Today, of course, the word teal is widely used to designate a number of other small ducks: Green-winged Teal and Blue-winged Teal are both vagrant species from North America, and I have seen Grey Teal and Chestnut Teal in Australia.

In the past, the Garganey, *Anas querquedula*, also had the word 'Teal' attached to its name, but that was dropped in the twentieth century. Its odd-looking English name comes via Gesner, who used a local Swiss word and referred also to an Italian form *garganello*. He suggested that the origin was onomatopoeic. The name is first seen here as *gargane* in 1668, picked up by Ray ten years later, and then formalised by Pennant. The specific form *querquedula* also appears to be based on the sound, but its usage goes back to two Romans, Varro and Columella. In his 1920 volume, Coward cites the name Cricket Teal as an alternative form based on the sound, but also refers to the Summer Teal as a reflection of its seasonal appearance. However, these names are mentioned by him only as curiosities.

Wigeon also has a long history, with the first use noted in 1513 as Wegyons, becoming Wigene in Turner 1555, and Widgeon in Merrett in 1667. That spelling still lingers, incorrectly, since Ray's version Wigeon became the definitive form adopted by Pennant in 1768. Ray also used an alternative name, Whewer, which was clearly based on the whistling call of the bird, as was Whistler, cited by Coward as a wildfowlers' name. The OED suggests that the first element of Wigeon (*wi–*) is rooted in an echo of the call. The scientific form, *Anas penelope*, has its roots in the Greek legend of the infant Penelope who was rescued from the sea by ducks. The Greek *pēnelops* was a sort of duck, so Penelope was named after her rescuers. Curiously, Linnaeus noted of the Wigeon, '*capite brunneo, fronte alba*' (brown head, white forehead), so it seems that he saw the crown as being white, rather than the 'creamy-yellow' of the *Collins Bird Guide*. That may have been a second-hand error, a trick of the light, or even a faded specimen, but was unlikely to be confusion with the American Wigeon, which he did not record.

Pintail is a very obvious name and is supported by the scientific name *Anas acuta* (sharply pointed). Other names existed before it was coined by Pennant in 1768. Ray used a local name, Cracker, but it was also referred to as the Sea Pheasant, possibly in confusion with the Long-tailed Duck. The Pintail once had the generic name *Dafila*, which was still used in Coward's 1926 list. The origin of the word seems uncertain, but it did attract Peter Scott, founder of the WWT, who named his daughter thus. Today it is the Northern Pintail in international terms, since other pintails exist. However, to me, this is the best of all ducks.

The (Northern) Shoveler, *Anas clypeata*, has another very obvious name, with its huge spatulate bill, but it was one 'stolen' from the Spoonbill by Ray in 1674 and standardised by Pennant in 1768. Indeed, Spoonbill is also cited by Coward as a local name for the Shoveler (which is preferred with the single L in the bird's name). The word *clypeata* is the Latin for 'shield-bearing', which is the image conjured by the massive bill being held in front of the bird. Interestingly, in Coward's 1926 list, it is still *Spatula clypeata*, which was perhaps a bit of an imagery-overkill. The Australian Shoveler, *Anas rhynchotis*, has a name which reflects Latham's hesitation about the bird's relationships, since the *–otis* element flags a curious 'likeness' to the Blue-winged Teal, *Anas discors*. That is puzzling, but his reasoning was apparently influenced by the similarities of the bill shape and the facial patterns of the Australian species to the American one. In fact both belong to the 'blue-winged' group of ducks, as do the Northern Shoveler and the Garganey, a fact which I find useful during return passage migration, when eclipse Garganey are more often found in company with Shovelers.

SHOVELER

The eighteenth-century English naturalist Mark Catesby originally recorded the Blue-winged Teal as *Anas variegata,* since the drake has a distinctive mixture of patterns and colours. That idea seems to be the root of the slightly obscure current name *Anas discors* (different). Curiously, though, the Latin version of 'blue-winged' appears in the name of the closely related American species, the Cinnamon Teal, *Anas cyanoptera,* whose similar blue wings contrast dramatically with the deep cinnamon body-colour. Though this is not on the British list, I was unable to miss an escaped bird as it flew into Titchfield Haven a couple of years ago.

Simply because their names are so curious, I cannot pass by two beautiful ducks which have arbitrary status in European records. Baikal Teal, *Anas formosa,* takes its English name from Lake Baikal in Siberia, while the scientific name means 'beautiful' – and among the duck species that is accolade indeed! The second is Falcated Duck, *Anas falcata,* and in this case both vernacular and scientific names link to the Latin *falx, falcis* (sickle), and refer to the gorgeous curved plumes worn by the drake (see also Glossy Ibis, Broad-billed Sandpiper and the falcons for other uses of that image).

The beautiful Mandarin Duck, *Aix galericulata,* is far too well established in Britain to deserve relegation to the supplementary pages of the *Collins Bird Guide,* since it is accepted by the BOU as a 'naturalised introduced species'. However, that firm status is relatively recent, since it did not appear in the 1923 list. While the magnificence of its plumage earned it the English name of a traditional Chinese civil servant, its generic name, *Aix,* is one of those anonymous water birds found in Aristotle. The specific term, *galericulata,* refers to the drake's superb mane, since it means 'be-wigged'. With those amazing orange sails on the wings, the naming of this species offered plenty of choice.

A small number of its only congener, the Wood Duck, *Aix sponsa,* live on the lakes near my home, still classified as 'escapes' (though there are enough nationally to warrant a page in the 2007–2011 BTO Atlas) and both these little beauties can be seen together. Beauty is reflected in the name *sponsa* (the bride) – 'prettily applied to this lovely duck', according to Coues. The only problem is that it is the male that wears the finery. This bird used to be called the Carolina Wood Duck, an indication of its American origins, while the former Australian Wood Duck, *Chenonetta jubata,* is now the Maned Duck, a description of the drake's short mane of plumage on the back of the neck. That is also reflected in the meaning of *jubata,* while the generic name speaks of a 'goose-like duck'.

Dive! Dive! Dive!

'Diving ducks' are a group which, in Britain at least, falls into two broad categories, the Aythyini and the Mergini. Most of the former prefer lakes and brackish habitats, but Scaup and Goldeneye seem to be equally at home on the sea. Conversely, of the more maritime Mergini, Goosander and Smew seem to have a penchant for inland waters.

Aythya is one of those cop-out Greek words taken from Aristotle and Hesychius to describe 'a sort of water bird'. Undoubtedly the exact meaning was lost in time and the word was simply 'available' in a pool of vaguely understood words.

Pochard, *Aythya ferina,* has one of the most intriguing names in the duck family. The word *ferina* simply means 'wild', as in our use of 'feral', but it should not be confused with the more precise use of Wild Duck, as in Mallard. Lockwood traces two distinct dialect words which lead to the name Pochard: Poker and Poacher. These words would appear to have a common root, however, and both suggest an action of 'poking about', a sense which is less obvious today in the second form. There is evidence that the word Poacher is attached to Wigeon in America, and it may well be that originally the two russet-headed British ducks were confused, since the term makes less sense with the feeding habits of a diving duck which spends much of the day asleep on the water. In keeping with a number of bird names, such as Mallard, the *–ard* suffix was attached, appearing first in Turner in 1544. Ray hesitated between Poker and Pochard, but Pennant, as so often, stabilised the name in the current form in 1768.

POCHARD

Ours should, of course be the Common Pochard in international terms, even though the two most likely confusion species do not bear the name Pochard. The two concerned are both on the British List as vagrants from North America: the Canvasback, *Aythya valisineria*, and the Redhead, *Aythya americana*. The first of these takes its vernacular name from the off-white gently vermiculated mantle and wings, which resemble the coarse fabric. The specific name is unusual in deriving from a favourite food of the birds, a species of aquatic wild celery which is in turn named after an Italian botanist, Antonio Vallisnieri (1661–1730). The drake Redhead earns its name from the fact that it has a more striking orange-red head than either the Common Pochard or the Canvasback. There is potential confusion in this name with the terminology of the sawbills, where juveniles and females are usually called 'red-heads' (see later), so this is one good example of where the use of a capital in a species name has an advantage.

Today we sometimes refer to the *Aythya* ducks in general as 'the pochards', though the Red-crested Pochard, *Netta rufina*, is atypical. *Netta* is the Greek for 'duck', while *rufina* is Medieval Latin for 'golden' or 'golden-red', the colour of the male's spectacular head.

The most widespread member of this group in my home county of Berkshire is the Tufted Duck, *Aythya fuligula*, which, in a design by Robert Gillmor, has long been the logo of the Berkshire Ornithological Club. The head tufts are more prominent on the drake, which also earns the title *fuligula* (sooty-throated).

Considering that the striking black plumage appears on all but the pure-white flanks and wingbars of the drake, this is a name of limited worth. Ray called it a Tufter, but Pennant saw to it that the current name became the norm. On the other hand the word 'Tufty' seems to fit very nicely in popular usage today. Coward cites an eye-catching name, Magpie-Diver, which was apparently used by 'sportsmen'.

TUFTED DUCK

The vagrant Ring-necked Duck is so similar at first glance that I once almost missed my first one in Canada by subliminally recording it as 'just a Tufty' as we drove past a lake. I had already seen a few in Britain, when I was impressed by what a feeble name this one has: the neck-ring is the last thing you notice, yet it is there in *Aythya collaris* too (and this one has typographically evolved into the Wring-necked Duck, which has of course been extinct a long time).

Lesser Scaup, *Aythya affinis*, became much more familiar to me in Florida than Greater Scaup, *A. marila*, has ever been at home. Thousands of the former winter on Tampa Bay, but the latter is a species which shows up for me occasionally, on the sea or on one of the local lakes. The word Scaup derives from a dialect word 'scalp' for mussels, on which it feeds. It was thus called the Mussel Duck or Scaup Duck, the latter name receiving the approval of Ray and then Pennant, but being abridged to the single (wildfowlers') usage of Scaup by Latham in 1797. Both forms were used until the 1920s, when Coward's 1926 list used the shorter form without an alternative.

The specific name *marila* means 'charcoal embers' or 'coal dust', while *affinis* implies 'an affinity to', in this case the Greater Scaup. I like to use memory-jogging words, and for this duck I find it expedient to make the link with 'refined', because the Lesser Scaup is certainly much slimmer head-on, with a distinctly narrower head, a point which Sibley makes clearly in his guide. Enough Lesser Scaup appear in Britain to make that a useful point.

On the other hand, it is worth remembering the dreaded words *Aythya* hybrid, since I recall that a huge number of people were wrongly claiming a Lesser Scaup one day at Welney. Some of the hybrids can be very similar to Lesser Scaup, so it is vital to consult Killian Mullarney's wonderful illustrations in the Collins guide before calling your friends.

Other confusions might arise with the scarce Ferruginous Duck, *Aythya nyroca*. While the common name tells us to look out for a 'rust-coloured' duck, I have always felt that a male in the sunlight is more like burnished bronze. The specific name is unusual in that it derives from *nyrok*, the Russian for duck. It was formerly known in Britain as the White-eyed Duck and appears as such in Coward's 1926 list. The name is obvious enough and is certainly easier to spell. (In modern birder-parlance, 'Fudge Duck' avoids any embarrassment about spelling.) One of the problems was that White-eyed Duck also appeared as the name of the similar Australian species, *Aythya australis*. That is a darker greyish russet in colour, a little more like the female Tufted Duck, as we saw in New South Wales. It now goes by the strange name of Hardhead, which was first noted in 1913, but of unknown origin. One of its Aboriginal names, *Bubbaloo*, deserves its place for its sheer curiosity.

No ducking the issue

Returning to my earlier comments about the way in which we can be divided from the North Americans by our common language, the story of the Long-tailed Duck needs to be given a special place, because in a way the Brits won that one. But before we crow, I must duck the issue by stating that we had nothing to do with it. The story goes that researchers needed to win the cooperation of Native American peoples in studying, for conservation purposes, the small duck, *Clangula hyemalis*, which was popularly known in America as the Oldsquaw. Since the term 'squaw' (a native Massachusetts word, according to Choate) had long been considered offensively racist in its application to Native American women, it was easier to negotiate the work by changing the formal vernacular name in 2000 to Long-tailed Duck, in keeping with the usage on this side of the Atlantic. In the circumstances it was probably better to lose a colourful name from the formal lists, though it is doubtful that it will disappear from popular usage for some time. Choate, incidentally, quotes the Cree name as *Hah-ha-way*, which he preferred. The decision to align with the BOU name probably coincided anyway with the work of the International Ornithological Union to standardise English bird names.

The word *Clangula* derives from the Latin *clangere* (to resound), an idea which is echoed in some of a number of traditional onomatopoeic names from the 'old country', which the Americans might have considered. In Orkney and Shetland the traditional name for the Long-tailed Duck was Calloo. In other areas of the north, Darcall and Coldie had similar origins. The odd name Coal-and-candlelight had currency in parts of Northumberland and Scotland, and it seems probable that it derived from a play on onomatopoeia with an added element to describe the black-and-white livery of the bird. I can just see some Northumbrian granny inventing that sort of fanciful nonsense for the grandchildren in a smoke-filled fisherman's cottage between singing verses of 'When the bwaot comes in'. Along with the Pintail, the species was also known in some parts of the country as the Sea Pheasant, clearly a reference to the drake's long tail-plumes.

However, the idea that a species like this could have popular names strikes me as mildly ironic: like many northern species, I have found it to be one of the most elusive of birds and can still only claim a few sightings, and all in recent years. This probably explains the second element of its scientific name, *hyemalis* (of the winter) since that is when we in the south have the best chance of seeing it. If I am envious of those who see such birds regularly, I look out at my own local birds and remember that they will be great rarities to someone else.

The ups and downs of the Eider

The Northumberland granny reminds me of a wonderful visit to the Farne Islands. The Farnes have the reputation of being the oldest nature reserve in Britain, and we have to thank St Cuthbert for that. Cuthbert was a monk who became Bishop of Lindisfarne in 676 CE and who created a retreat on Inner Farne, where he built a chapel which still stands among the nesting birds. The bird which we know today as the Common Eider, *Somateria mollissima*, was one of the species which nested safely on the island. It became known locally as the Cuthbert Duck, and other corruptions of that idea became such as Cuddy Duck and Culver Duck.

The word Eider was first used by the Danish naturalist Worm in 1655. He had taken it from an Icelandic form, æður (or æðarfugl), which in turn came from the Old Norse and meant 'down-bird', referring specifically to the female. Another folk-name, Colk, used in the Western Isles, conveys a Gaelic version of the same idea. As a bird of northerly latitudes, the female produces copious amounts of down from its breast in order to keep the eggs and young warm. Humans had long harvested the down to line clothing and bedding. The eiderdown, a well-stuffed coverlet of down, remained in widespread use well into the twentieth century: I certainly snuggled gratefully under one on many a winter's night as a youngster. The continental duvet replaced it, but retains the link, since *duvet* is the French for feather-down.

Pennant was the first English naturalist to adopt the word Eider, but he added 'Duck', which has now been formally dropped, with eider serving as the generic form for four species: Common Eider, King Eider, Spectacled Eider, and the less closely related Steller's Eider.

The scientific name of Common Eider, *Somateria mollissima*, records the importance of the down, but in a curious way: it is composed of the Greek *sōma, somatos* (body) and *erion* (wool), plus the Latin for 'very soft'. Now that is what I call woolly thinking! Surely there was a better word than *erion*?

FEMALE COMMON EIDER

An extant Shetland name for the bird is the Dunter, which is apparently derived from a verb of Norse origin, to dunt, meaning to bob up and down in the waves. There will be a good many like me who have tried to focus on distant sea-ducks in the waves and have called them many things, but the word Dunter never occurred to me. In truth, I find Eiders so much easier to see than the Long-tailed Duck, since I can now almost guarantee to find them on the Solent and have found them more than once during the summer moult on shores as far apart as Essex and Nova Scotia.

However, one bird stands out in my memory above all others – a female on her nest in a thick, lush bed of soft weeds, just yards from St Cuthbert's Chapel on Inner Farne – just as her ancestors would have been in the seventh century.

Just how black can a black duck get?

Our most common sea-duck will fit that bill very nicely, especially in the light of the alternative names, Black Duck and Black Scoter –'blacker than any other duck', according to Coward. Common Scoter is *Melanitta nigra*, which consists of the Greek *mela* (black), *nētta* (duck) and the Latin *nigra* (black). By which point we have the picture!

But the real interest here is the word 'scoter', especially since major dictionaries like OED and Merriam-Webster duck the issue by classifying the word as 'of unknown origin'. Lockwood may have been seduced by the concept of 'blackness', since he suggested that the name arose from an error of transcription of the word 'sooter', a speculative form which would relate to a German name of similar meaning for the species, and one which would fit the colour of the drake.

COMMON SCOTER
(REPRODUCED BY PERMISSION OF ERNEST LEAHY AND BBAG)

The American, Choate, on the other hand, leads me on a more convincing trail. Firstly he links the word with Coot, 'which is used by hunters for the scoters'. Indeed, a modern website for American duck-hunters confirms the link, by quoting the names Sea-coot and Scooter as alternative names for Scoter. On the face of it it seems that an old usage travelled to America, but has been long forgotten in Britain and overlooked by lexicographers. The word, according to Lockwood, originates from Yorkshire, and was first used formally by Ray in 1674, by which date the colonisation of America was already under way. This explanation looks suspiciously simple, but a closer look at Choate brings a further undeveloped lead, for he cites MacLeod's evidence that '*macreuse* in the north of France means a scoter, and in the south a coot'. In fact, the etymology of that word is directly linked with an old French name for Coot, *macrolle*, and thence to a Dutch dialect form *meerkot*, meaning 'sea-coot'. Further evidence comes from 1555, when Pierre Belon, in *L'histoire de la nature des oyseaux*, noted the name *macroule 'diable de mer'* (meaning 'sea devil'). Significantly, Lockwood has Devil as one of the old names for Coot, and quotes a 1580 record which states that 'because of its blacknesse is called a *Diuell*'. Given all that evidence, it seems most probable that Sea-coot is the origin for our mysterious word Scoter. From my own experience, a Common Scoter was very hard to spot among Coots on a local gravel pit since they are of similar size, colour and shape.

The word *fusca* (very dark) is used as an alternative form of blackness in Velvet Scoter, *Melanitta fusca*. The name Velvet Duck was proposed by Ray in 1678: 'the feathers of the whole body are so soft and delicate... ', and used by Pennant, but it was Fleming who used the name Velvet Scoter in 1828.

The American species, the Surf Scoter, has long been known as a rare vagrant to British waters, but in recent years two new species have been created on the American side, in the decision to split the Back Scoter, *Melanitta americana*, from Common, and the White-winged Scoter, *M. deglandi*, from Velvet, thus bringing new names into the field guides and the possibility of five black duck species. The nineteenth-century French ornithologist C. D. Degland is commemorated in the second new name.

And just to leave this dark section with at least some colour, there is another species very rare in British waters, the oddly named Harlequin Duck, *Histrionicus histrionicus*. In the traditional Commedia dell' Arte theatre, Harlequin wore a multi-coloured suit. From this comes the concept of *Histrionicus* (theatrical). (*Exeunt scoters, stage right*)

Saws at sea

The mergansers are characterised by tooth-like serrations along the edge of the bill, which earn them the familiar collective name of 'sawbills'. This adaptation helps them to catch and hold slippery fish.

The largest of these is the Goosander, *Mergus merganser*. Goosander may or may not be derived from the word goose, but one logical possibility is that it includes goose plus *ende* (duck). Lockwood, on the other hand, is not convinced of that origin. Prior to Ray it was Gossander, but he, and later Pennant, used the current spelling.

In spite of its superficially similar structure, the word 'merganser' places the goose element, *anser*, behind the Latin word *merg* (diving). It seems that the word 'goosander' resulted from a natural evolution, while 'merganser' is a later artificial scholarly creation. The Americans, of course have no truck with that distinction. The larger species is called the Common Merganser, in line with Red-breasted Merganser, *Mergus serrator*, which is the full name of the one seen in Britain. *Mergus* was an unidentified seabird mentioned by Pliny, Varro and Flaccus. The serrated bill gives rise to the specific name *serrator*, meaning 'sawyer'.

The much smaller Smew, *Mergellus albellus*, was once known as the White Nun, which is ironic given that only the male is white. That was Ray's preferred name. In 1668 Charleton referred to it as Diving Widgeon, but it was, as usual, Pennant who endorsed the form we now use. Given similar forms in Dutch and German, Lockwood feels that the word Smew has much older roots. The OED implies that there is a link with 'small'. That element is emphasised by the fact that *Mergellus* is a diminutive of *Mergus*, while *albellus* does the same for *alba* (white).

The Common Goldeneye, *Bucephala clangula*, holds no mystery in the vernacular form: the name was actually coined by Ray and endorsed by Pennant. *Bucephala*, on the other hand, is a bit more colourful, since it derives from the Greek *bous* (ox) and *kephalē* (head), in recognition of the large-headed appearance of the bird. The word relates to the name of the American duck, the Bufflehead, and of course to buffalo. The specific name *clangula*, from the Latin *clangere* (to resound), may well relate to the whistling sound produced by the wings, since the voice is quite insignificant compared to that of the Long-tailed Duck, which has the same word as its generic name. The Bufflehead, *B. albeola*, has a specific name to reflect its colour and small size, since *albeola* is another diminutive of *alba* (white). Another close relative is Barrow's Goldeneye, *B. islandica*. While its specific name reflects its geographical base, the common name was given by Swainson to honour the founder of the Royal Geographical Society, Sir John Barrow.

Both of the latter birds appear as vagrants in Britain, as does the Hooded Merganser, *Lophodytes cucullatus*, which also introduces a new genus meaning 'crested diver'. The 'hood' is a dramatic black and white crest, which the male raises in display. That feature is also represented by its Late Latin specific name.

In all of these species, the females are distinctly different from the striking males and have variants on the theme of grey or greyish bodies with russet or brown heads. Immature males are similar. As a consequence of this it has become the norm to refer to that plumage form as a 'red-head'. (A similar use of language occurs in the term 'ring-tail' for female and young harriers.)

Stiff-tails

Here we enter into a totally different group of diving ducks. Strictly speaking, these would not be of great interest to the average British birder had it not been for the escapes of the North American Ruddy Duck, *Oxyura jamaicensis*, from Peter Scott's Slimbridge collection, and their subsequent proliferation. It is ironic that a man dedicated to the conservation of species, and almost single-handedly responsible for saving the Nene or Hawaiian Goose from extinction, should have been indirectly the source of a threat to another endangered species. The controversy surrounding the recent British cull of the attractive little Ruddy Ducks was bitter, and the work expensive and brutal. In 1980, Peter Scott wrote: 'It seems doubtful to me that the Ruddies will ever become a pest or drive out any other species.' At that stage there was no evidence of its drift towards Spain, and it would have been interesting to record his perspective on the issue twenty-five years later, but he died before the cull became a reality.

Oxyura means 'sharp-pointed tail', but these ducks are generally referred to as 'stiff-tails'. The fear of this American intruder damaging the genetic line of the heavily endangered White-headed Duck, *Oxyura leucocephalus*, in Spain has been well documented. There is nothing unexpected in the specific name of the latter, which translates the English version. We encountered this species in Andalucia, and we also saw a single specimen of *Oxyura australis*, the Blue-billed Duck, at Lithgow in New South Wales. I have to confess that I rather miss the little American intruders.

Give a little whistle

With many escapes from wildfowl collections, it is worth adding a word about the subtropical whistling ducks, which represent a further subgroup of wildfowl. There are eight species in all. Having seen two in Australia and one in Trinidad, I was particularly surprised to see that an escaped Fulvous Whistling Duck, *Dendrocygna bicolor*, survived among the wild ducks at Titchfield Haven for a year or two before disappearing. The generic name means literally 'tree-swan', a reflection of their habit of tree-perching on the one hand and of their long necks on the other. The common name Fulvous in this case refers to a fawn-brown body plumage, which is two-toned (*bicolor*).

WHOSE GAME IS IT? PHEASANTS, PARTRIDGES AND GROUSE

First, a bit of grousing... My feelings towards the general term 'game birds' are rather akin to my discomfort about 'wildfowl'. There will be those who consider this to be over-sensitive, but I cannot feel comfortable about the idea of shooting live creatures for fun and calling it 'sport'. The word 'game' is an associated idea, and one which I would happily replace with something more worthy.

RED-LEGGED PARTRIDGES

I dealt with the Common Pheasant earlier, and in a way it is an appropriate separation, since it was originally a non-native species. That is true also of the less-widespread Golden Pheasant, *Chrysolophus pictus*, and Lady Amherst's Pheasant, *C. amherstiae*. The generic name of both means 'golden crested', but the former is *pictus* (painted), while the latter, of course, commemorates Sarah, Countess Amherst (1762–1838).

The Red-legged Partridge was also an introduction, in this case by Charles II following his restoration to the throne in 1660, after a long exile in France. The dominant name in my young days was still French Partridge. Curiously, I have seen 'proper' wild ones only rarely in France, and just once in Andalucia. Their scientific name, *Alectoris rufa*, hints at their semi-domesticity, since it means 'rufous chicken'!

A few decades ago, a number of their more easterly relative, the Chukar, *Alectoris chukar* (from the Hindi word *chukor*), were released in this country, but that practice was stopped because they out-competed the wild Red-legged, which had been here for so long. As it is, the Game & Wildlife Conservation Trust (GWCT) states that six million Red-legged Partridges are released each year – which is about the same as the entire population of Woodpigeons. The Pheasant population is sustained in a similar way.

A third species, the Rock Partridge, *Alectoris graeca* (Greek), is the indirect star – as *la bartavelle* – of Marcel Pagnol's autobiographical story, 'La Gloire de Mon Père'. In that story of his bookish schoolmaster father's first outing with a shotgun, the Rock Partridge is *le roi des gibiers* (king of the game birds) and a target profoundly revered by the locals. Young Marcel accidentally flushes a pair and the father accidentally shoots both in a lucky reflex. They keep mum about the fluke and his father immodestly bathes in the glory of the legend he has become.

The native Grey Partridge, *Perdix perdix*, used to be our Common Partridge, but modern farming practices have made it far from common these days. The word Partridge is a direct evolution from the Greek and Latin names for the species, *Perdix*, which relates to the concept of camouflage (i.e. getting lost). The idea is found in the modern French verb *perdre* (to lose), and the lost daughter in Shakespeare's *Winter's Tale* is Perdita. Medieval French, of course, was the route by which the word travelled into English as *perdriz*, corrupting into *partrich* by 1290, and then appearing in the present spelling in 1579.

A small cousin of the partridges is the migratory Quail, *Coturnix coturnix*. The word *Coturnix* is the Latin for 'quail', and was used by Gesner in 1555. Unusually the vernacular name, which comes to us from Norman French as *quaille*, developed from a Franconian German word, rather than from Latin. Lockwood points out that it starts out as an imitative word and evolves with a diminutive ending to become the root of today's form. That genesis is far less obscure than the birds themselves: I have stood within metres of them as they moved around calling in minimal cover, and still never seen so much as a movement. In international terms we need to refer to it as the Common Quail: in fact we were lucky enough to see its close relative the Brown Quail, *Coturnix ypsilophora*, as we left our Kakadu campsite. The strange specific name of the Australian species means 'bearing the letter Y' and refers to the dark markings on the bird's flanks. Painted Buttonquail, *Turnix varius*, which we saw in the Blue Mountains, belongs to an entirely different family, though an external similarity links it in both vernacular and generic name to the true quails.

All of the other British game birds were well out of my natural geographic range for many years, but I managed eventually to catch up with the two grouses (or should that be grice?). In fact, Lockwood suggests that the word 'grouse' (also once grous) is technically a plural form, citing evidence in earlier spellings of its use alongside the plural form of pigeons. He postulates a singular, *grow*, which he links to a French form *grue*, based on a Late Latin form *gruta*, which is evidenced in Giraldus, *c.*1135, as meaning 'field hen'.

The status of Red Grouse, *Lagopus lagopus scotica*, has long been disputed. It is considered by some authorities to be a subspecies of the Willow Ptarmigan, *Lagopus lagopus*, though Madge and McGowan (2002) assert that it is a separate species, *Lagopus scotica*. What is clear is that Willow Ptarmigan and Red Grouse use totally different habitats, and that the British birds do not have any white plumage at any stage of the year. A further complication is the current preference for the name Willow Ptarmigan over the alternative Willow Grouse. Clearly that move brings it into line with its close relative of high altitudes/latitudes, which was formerly known as *the* Ptarmigan and is now modified to Rock Ptarmigan, *Lagopus muta*.

Aptly, with a specific name *muta*, meaning 'silent', the Ptarmigan had no say in the matter. The irony of that is found in Lockwood's explanation that the word Ptarmigan derives from the Gaelic rendering of the bird's hoarse croak to become *Tarmachan*. It was Latinised as *Terminganis* at the end of the sixteenth century and morphed into a pseudo-Greek form in its current spelling by Sibbald in 1684, then standardised as a result of its adoption by Pennant in 1768. *Lagopus* means 'having feet like a hare', and refers to the heavily feathered legs of all three of these birds (it occurs again in the Rough-legged Buzzard, *Buteo lagopus*).

Black Grouse is still classified as *Tetrao tetrix* in the 2015 BOU list, though it is now *Lyrurus tetrix* in the IOC list. Clearly it is is less closely related to the first three (or should that be two?). It was known also as Black Game and Heathcock until Yarrell standardised it in 1843. *Lyrurus* means 'lyre-tailed', which is a reference to the curved tail-feathers of the male. The word *tetrix* comes from Aristotle and refers to a species of ground-nesting bird. The same word appears in a modified form in the generic name of small relative, *Tetrastes bonasia*, the Hazel Grouse, which was long known as the Hazel Hen. This sedentary bird of northern and eastern Europe is sometimes classified in the genus *Bonasa*, which, like the specific name, seems to be of obscure origin. The concept of a 'maned bull' seems to fit the North American Ruffed Grouse, but the name was first applied to the Eurasian species, which lacks a ruff.

Capercaillie, *Tetrao urogallus*, might seem to be unique, but with the Black-billed Capercaillie in eastern Asia, this one sometimes needs the addition of Western in front of its name. The vernacular name is of Scottish Gaelic origin and started life as *capull coille* (horse of the woods), because of its song. Before Gould adopted the form Capercailzie in 1834, it was known by English naturalists as the Wood Grouse. Once again Yarrell standardised the spelling in the current form in 1843.The generic name *Tetrao* is clearly related to *tetrix*, while *urogallus* possibly has German roots in *Auerhahn*/*Auerhenne*, via Gesner (1555). Those forms literally suggest a 'long-tailed fowl', as in the Greek *oura* (tail). The evidence of other scientific names beginning with *Uro–* tends to support that idea.

LOONY TUNES: DIVERS OR LOONS?

I met my first winter-plumaged Common Loons on Tampa Bay in Florida, but had already seen a number in magnificent breeding plumage in Nova Scotia, where they flew past our apartment window at night uttering their eerie wailing call, which always says *On Golden Pond* to me. In short, I was happy to think of them as 'loons' on that side of the Atlantic. The word 'loon' actually travelled from Britain to America with the settlers. In Britain it was more frequently applicable to what we now call the Red-Throated Diver. It was a Northern and East Anglian form which may have had roots in the Old Norse *lómr*, where its sense refers to the sound of moaning. The transition to loon may have been helped by the Latin roots of *lunatic* (originally 'one suffering from moon-madness').

These North American encounters were all very well, but as a child I was profoundly influenced in my love of nature by such authors as Arthur Ransome. Like the Swallows and Amazons, I did a bit of sailing, experienced the rip-tide which featured in *We Didn't Mean to Go to Sea* (in that case the harbourmaster's launch spotted our problem in time), and grew up within a few kilometres of the real Peewit Island. In short, their story in *Great Northern?* introduced me to the Great Northern Diver. I am an old romantic in these matters, I suppose, and therefore cannot describe just how important the sighting of my first Great Northern Diver was. I had waited well over fifty years since reading the book and had seen plenty of Common Loons in North America, but one January, at the start of the twenty-first century, I at last caught up with an authentic Great Northern Diver on a Berkshire gravel pit. Both were, of course, *Gavia immer*, so it was only the electric current of childhood dreams that gave the British find its real significance.

COMMON LOON OR GREAT NORTHERN DIVER

In many ways that story is probably the one which really underpins the motivation behind this work. Do these things really matter? I think they must, because I was quite uncomfortable with the hybrid, Great Northern Loon, which is the name used in the second edition of the *Collins Bird Guide* in accordance with the then IOC usage. I felt that the compromise was mere fence-sitting, even though reason reminded me that the word 'loon' was actually coming back to its origins in Britain. However, version 6.1 of the IOC list (January 2016) now shows that the name has been discarded and replaced by Common Loon. That move may be decisive, but from a British perspective the decision turns reality on its head, since it is by far the rarest of the three British breeding species.

The very name Great Northern underlines the fact that it is a much more northerly species than the other two, while the specific name *immer* derives from the Norwegian name for it. Also there is the archaic term Ember Goose, which comes from the Faroe Islands, where the bird appeared in the Ember Days, just before Christmas. For its sheer colour alone I must include here the obscure Hebridean name *bunivochil,* which is derived from a Gaelic word for a herdsman, whose call to the cattle is echoed by the wild call of the bird.

Unfortunately there is no colour at all in the generic name *Gavia*, which is Latin for an 'unidentified seabird', but the Red-throated Diver/Loon is *Gavia stellata*, which recalls the scientific name of the Great Bittern (*Botaurus stellaris*). In the case of the diver the 'starry pattern' (*stellata*) refers to the 'frosted' winter plumage, in which white feather-margins and other spots make this a much paler bird than the other divers. The common name is only of use for identifying the bird in summer plumage, as is also the case with the Black-throated Diver/Loon, *Gavia arctica*. And that reminds me that the *Collins Bird Guide* offers a truly puzzling compromise when it gives the Black-throated Loon the alternative name Arctic Diver, presumably to marry the North American Arctic Loon with the British Black-throated Diver. The formal names on the IOC list do not include that option.

RED-THROATED DIVER

Another name which illustrates only too well the parallel thinking of two different English-speaking nations is that of the largest of the family, *Gavia adamsii*. The specific name here commemorates the British surgeon-explorer Edward Adams (1824–1856), a member of the expedition which went in search of Franklin. However it is the English name which is fascinating, because it grew as White-billed Diver on this side of the Atlantic (and appears as White-billed *Northern* Diver in the BOU 1923 list and Coward's 1926 list). However, it is the Yellow-billed Loon on the American continent and in the IOC list. In truth the bill is neither white nor yellow, being of a pale ivory-cream shade. Ivory-billed Loon/Diver would have been suitably neutral.

In the next section I will be discussing issues behind the name *Colymbus*, by which the loons were formerly known: in that are further elements of disagreement between the two traditions. In general, in the naming of the divers, and as later discussed in the naming of the skuas, the strained efforts of the international body to impose some sort of compromise between two divergent branches of the same language, and between two different cultures, seem to run into a 'Cinderella's slipper' problem: compromises can be extremely ugly. On the other hand *The Rite of Spring* was once booed off the stage because the Parisian public did not appreciate that particular musical language – and now it is a classic. Maybe the compromise will grow on a new generation, but for the time being I feel that the current solutions are far from perfect.

A TRAGEDY OF FASHION AND A COMEDY OF ERRORS: GREBES

Grebes are linked to the Red-throated Diver by one of their colloquial names, Arsefoot. This was a term used in the sixteenth and seventeenth centuries for the grebes and some divers, all of which have their propeller mechanisms set close to the back end. It is reflected in the later scientific name for the grebe family, the Podicipedidae. The first element, from the Latin word *podex, podicis*, means 'vent' and the remainder is derived from *pes* (foot). The term was devised by Latham in 1787 to replace the form *Colymbus*, which was used earlier and appears in Linnaeus.

The older name, which was used by Gesner in the sixteenth century, eventually disappeared, partly because it was considered to be too easily confused with *Columba* (pigeon/dove), but also because it became embroiled in Anglo-American disagreements. In the *Systema Naturae* it still applied to the divers as well as the grebes. The Americans continued to use it for the former category in spite of later protests from the BOU: Jobling quotes a 1915 statement of that sort. It was finally suppressed by the ICZN in 1956. However, the American Pied-billed Grebe – a not-infrequent vagrant to Britain – still sports a hybridised generic name in the form *Podilymbus*.

GREAT CRESTED GREBE

The word 'grebe' came into English from French via Brisson, when Pennant adopted it in 1760. The same Pennant had also preferred the French-rooted word Guillemot, and this is another example of the sort of pretentiousness which made French seem more acceptable than basic Anglo-Saxon – a phenomenon which has still not gone away. The word *grèbe* had been used for 200 years before that in France and is considered to be rooted in a Savoyard name for a gull species. At the time of its transference to English, it was known in the fashion trade, where grebe skins were used to decorate garments. (Later the fashion industry created whole tippet cloaks of grebe feathers, which, together with the exploitation of egret plumes, was driving such birds towards extinction. This provoked an outrage among a group of nineteenth-century women and led to the formation of the RSPB.) The Great Crested Grebe, *Podiceps cristatus* (crested), was a principal victim of the trade and its very survival in Europe was threatened. It is once again a relatively common species.

LITTLE GREBE

As for the smaller grebes, the Little Grebe, *Tachybaptus ruficollis*, was commonly known as Dabchick (or a number of regional variants thereof, including Dobchick). This is one of the few traditional names which is still widely known and used. The concept of 'dab' probably relates to the idea of a small quantity (as in 'a dab of paint'), while 'chick' is clearly reinforcing the smallness of the bird.

The word *Tachybaptus* means 'to sink fast', and I think that the concept would have puzzled me if I had not seen the escape behaviour of the very similar Australian species, *Tachybaptus novaehollandiae*, which, if surprised in an open pool, *sinks* rather than dives, leaving virtually no ripple. The art of finding them then is to scour the margins, where they will pop up among the vegetation. That was a useful way of separating them in a brief view from the Hoary-headed Grebe, *Poliocephalus poliocephalus*, which scuttles off over the surface. This latter name derives from the Greek *polios* (grey) and *kephalos* (headed).

The problem with the specific name of the Little Grebe, *ruficollis*, is that it means 'red-necked', and we already have the Red-necked Grebe, *Podiceps grisegena*, which means 'grey-cheeked'. Both of those names are accurate in the breeding season, but they do muddy the waters a little.

It gets more complicated when you travel to North America. In some ways the security of the scientific names is helpful, because two of our smaller grebes take on new names over there. The Black-necked Grebe, *Podiceps nigricollis*, becomes the Eared Grebe, while Slavonian Grebe, *Podiceps auritus*, becomes the Horned Grebe. The golden head-trimmings of the two species are responsible for those names, which are perfectly reasonable in English. In fact, the names Horned and Eared Grebes were created by the English naturalist and artist George Edwards (1693–1773) and Pennant put the two names into general use in 1768. The adoption of these forms in America was akin to that in Britain, where Pennant's two names were in wide use until 1912, when the *Handlist of British Birds* was published by Hartert *et al.*

RED-NECKED GREBE

All of that seems very reasonable until you consider that the *auritus* of the Horned Grebe means 'eared'. In fact, something seems have gone wrong between times, with the confusion originating in the *Systema Naturae* of Linnaeus. To explain this better it is necessary to understand that each species entry by Linnaeus is accompanied by notes showing the alternative names and their origins. Thus, under *Colymbus*, we see the heading *auritus*, and among the source notes there are 'Eared Dobchick, Edw[ards]', and 'Eared Grebe, Lath[am].' Clearly, the words *auritus* and 'eared' are correctly linked at that point. On the following page, an entry under *C. cornutus* states 'Eared or horned Dobchick' against Edwards' name, while the name Horned is noted as Latham's use. In very simple terms, there was no confusion at this point, since several other descriptions of each species support the fact that Linnaeus had his data well ordered. The most likely explanation for the subsequent transference of *auritus* to the wrong species seems to be rooted in the ambivalence of the second note attributed to Edwards. What remains is that the name *Podiceps auritus* is attributed to Linnaeus in 1758, and attached to Horned (Slavonian) Grebe, but exactly how that detail slipped from one species to the other is a mystery. And, by the way, *cornutus* means 'horned' and is now the trinomen of the North American subspecies, *P. a. cornutus.*

SLAVONIAN (HORNED) GREBE

The trail leading to the two names currently used in Britain begins later than Pennant. The word 'Slavonian' was used in 1802 by Montagu, who referred to a comment by Latham that the bird was known in Sclavonia (an area of northern Russia). The name was adopted by Jenyns in 1835, then Yarrell in 1843, and formalised by the BOU in 1883. The current spelling was used by Hartert *et al.* in 1912. It was as recently as the Hartert list that the English translation of the specific name *nigricollis* came into being as Black-necked Grebe and was adopted formally: it may well be that the intention was to avoid the confusion now inherent in the names of Eared Grebe and Horned Grebe. There was clearly some concern about the misplaced ears, since Coward commented in 1923 that 'the B.O.U Committee make a great mistake in trying to make the English name of the Slavonian Grebe agree with the specific name.' Today the IOC list offers a compromise solution, with the preferred names being Horned Grebe and Black-necked Grebe. Even so, the 2014 second edition of Sibley has retained 'Eared', and it is likely that 'Slavonian' will continue to be used in Britain.

This story illustrates why my interest in this area began in North America: while we sometimes appear to be divided by the same language, the truth lies in the cultural choices made at different historical points.

And are names really important? I think so. To me the Slavonian Grebe is a small, dark-and-light bird seen distantly in cold winter seas, where their monochrome shading makes them hard to find. I associate them with cold fingers and with patient hours of scanning through the scope during that masochistic subset of birding called 'sea-watching'. The Horned Grebe, however, takes me back to a sunny winter Florida and to dawn visits to a jetty on Tampa Bay. That part of the bay held many wintering birds and nothing seemed the same from day to day. I frequently observed flocks of up to twenty-eight Horned Grebes on the water, diving with uncanny unison. An extract from my memoir on those trips tells clearly why anyone who misses the dawn misses magic – I have long been an early bird:

The sheer beauty of some of the dawn light settings that framed those grebes on the bay will remain among the most memorable moments. With dawn glowing red on a taffeta-blue surface, the birds would show as starkly black above and white below, an inverted replica of each doubling the size of the image in the scope. When all had dived in their formation ballet, the water would be still, as if nothing was abroad in the dawn. Then, one by one they would twinkle to the surface, the droplets on their white feathers flashing like tiny diamonds.

At moments like that, I didn't really care about the confusion in their names.

THE PATTER OF TINY FEET: SHEARWATERS AND PETRELS

Shearwaters and petrels all belong to a broad group of highly pelagic species, which includes such as storm petrels and albatrosses. These are the Procellariiformes (a word meaning 'storm gulls'), which are characterised by a special adaptation of the nostrils, which form tubes along the top of the bill. From this they are often referred to as 'tubenoses'. The device helps to drain salt excretions (which are filtered by salt glands in the head) into a channel on the bill and away from the eyes.

To an average birdwatcher most of this group remain something of a mystery, simply because it takes a special effort to see any of them at all. The most familiar is the Fulmar, *Fulmarus glacialis*, since the species did oblige British birdwatchers by colonising our shores during the twentieth century. As the specific name suggests, this was once considered to be a bird of the frozen north until it changed its habits and spread south. The unattractive behaviour of the Fulmar led the Viking colonisers of St Kilda to name it with the form we still use, meaning 'foul gull'. In simple terms, the nesting bird and its young defend themselves with a jet of regurgitated smelly oil, which clogs the feathers of a predatory falcon or gull, possibly with fatal results – a practice which perhaps ensured that the young were less collected for food than those of the Gannet.

The Fulmar is a 'petrel'. That word was added to the name by Yarrell in 1843 and retained in Hartert *et al.* in 1912, but it is now rarely applied. According to Lockwood, that word has its origins in the 'pitterel' (pitter-patter) of the feet of smaller species on the water. Choate quotes the OED for evidence that the current spelling was first used by the mariner William Dampier in 1703, while Lockwood feels that subsequent links with the name of St Peter, and walking on water, are purely coincidental.

The Blue Fulmar is the name given to a darker, grey morph, which mostly occurs further north. The word Mallemuck is cited by Lockwood as being a Scottish dialect name for the Fulmar, which appears to have been transferred there by Dutch sailors. The similar word Mollymawk is found in other parts of the world as a name for albatross species. Both words mean 'foolish bird' in the original Dutch (cf. booby and *Morus*).

Others of its relatives are much harder to see: I recall that *Puffinus puffinus* (Manx Shearwater) showed up on our way back to shore from the Farne Islands. Just the one, but close enough for those of us still alert to get a good water-level view, as opposed to the usual one from a coastal vantage point. That way we could really see it shearing the water on its stiff wings. The flight action described by the word 'shearwater', in which the wing-tips sometimes appear to cut the crests of waves, has evolved in this family as a means of travelling huge distances with great economy by using the energy of wave-generated air movement.

Manx Shearwaters used to nest on the Calf of Man, hence the name, but the colony was wiped out by rats by about 1800. Today they are found on rat-free islands such as Skomer, Skokholm and Rum. The scientific name is a curio, because it originally was a Middle English word (*Poffin* or *Pophyn*) created to describe the cured, fat nestlings of the shearwaters, which until the eighteenth century were gathered as a food delicacy. Presumably the word shares its value with 'puffed out' and 'puff pastry'. Suffice it to say that fat baby birds suffered in other cultures too, and the topic will return later.

Finding other shearwaters is a matter of luck and location. Sooty Shearwater, *Ardenna* (*Puffinus*) *griseus*, showed up while we were whale-watching off Halifax, Nova Scotia. Its dark brownish-grey colours are reflected in both the English and the scientific names, but the interest of that one was that it was 'wintering' in our northern-hemisphere summer, and a long way from its breeding territory. Here it is worth mentioning that the IOC Master List now places some of the former *Puffinus* species in the genus *Ardenna*, while they are as yet unchanged in the BOU list.

In 2011, we managed to see Balearic Shearwater, *Puffinus mauretanicus*, off Andalucia with the former Mauretania (Morocco) opposite and the Balearic Islands relatively nearby. This species and its close relatives are better understood than a few years ago. Splits have occurred, so our reference books have sprouted new names, some of which are intriguing. My favourite is Yelkouan Shearwater, which must be the only English-language bird name originating in Turkish. It means 'wind-chaser'. Very recently, the difficult name Macaronesian Shearwater (taken from the collective term for the Azores, Canaries, Madeira and Cape Verde) has been sunk by taxonomic changes which have reversed the former lumping of two species: Barolo Shearwater, *Puffinus baroli*, and Boyd's Shearwater, *Puffinus boydi*. The former commemorates the Italian philanthropist, the Marchesi Carlo di Barolo (1782–1838), whose name also lives on in the wines from his estates, while the latter was Lieutenant Boyd Alexander, an explorer and collector who died in 1910.

On the day we saw the Balearic Shearwaters, I also became aware for the first time of Scopoli's Shearwater, *Calonectris diomedea*, which was further offshore. The story of that bird is complicated, and to trace its vagaries is worthy of a volume in itself. This is the essence of the trail, as succinctly as I can express it. The Mediterranean bird was described by Giovanni Scopoli in 1769, but he apparently considered it to be a subspecies of Great Shearwater. From Atlantic specimens, the American Charles B. Cory established in 1881 that it was a species in its own right. Subsequently, the BOU list, as used by Coward in 1926, shows that Scopoli's bird was known here as the Mediterranean Great Shearwater, *Puffinus kuhlii kuhlii*, with Cory's as the Atlantic Great Shearwater, *P. k. borealis*. The Great Shearwater was listed alongside as *Puffinus gravis* (now *Ardenna gravis*), so any confusion now existed only in the English forms. Subsequently the first two were reclassified as a single species, *Calonectris diomedea*, and Cory's Shearwater became the commonly used vernacular form, since that was the one seen in British waters. More recently, the name Scopoli's Shearwater has been resurrected and appears in the second edition of the *Collins Guide*, since it is possible to separate it from Cory's visually. A

2012 report in the BOU journal, *Ibis*, recommended a split, and that is confirmed in the current IOC list. Scopoli's is now *C. diomedea*, and Cory's is *C. borealis*, the latter being further split from *C. edwardsii*, Cape Verde Shearwater. As we saw it, Scopoli's is a large shearwater, larger even than the Fulmar, a fact which is recorded in the meaning of the current generic name, *Calonectris*, meaning 'noble petrel'. As for *kuhlii*, Kuhl was one of Temminck's assistants and, I can only presume, did more work on Scopoli's bird. Even if he has now been lost from the name of this species, he is listed by Beolens & Watkins for links to other species. That convoluted tale is, in many ways, symbolic of the elusiveness and mystery of such pelagic species, even in the twenty-first century.

The very smallest members of this group are known as storm petrels. In British waters the most common one is the tiny European Storm Petrel, *Hydrobates pelagicus*. Storm Petrel was probably a mariners' name, which was adopted as Stormy Petrel by Pennant in 1776 and modified to the current form by Jenyns in 1835. Its alternative name, Mother Cary's Chicken, like the definitive name, was based on a superstition that the birds presaged a storm. Choate quotes Yarrell, who linked it to a reported 'sea-hag' who in another legend was an aunt to Davy Jones. The latter's locker was a metaphor for the ocean depths, which held the remains of drowned mariners. Perhaps the most convincing of all is that it was a corruption of *madre cara* (dear mother), a phrase of prayer uttered by southern European sailors as storms approached. Lockwood speculates that 'Cary' is a corruption of Mary, which also implies that imperilled Catholic sailors would have been praying hard. There is in fact every cause to feel that all these elements would link together quite smoothly, with the prayer before the storm, a misunderstood foreign phrase transferring into an English name, and a link of that name to traditional maritime superstitions. The species generic name *Hydrobates* is close to the origins of Petrel, since it means 'water-walker', while *pelagicus* means 'of the sea'.

There are a number of other Atlantic species, some of which appear in British waters.

Leach's Storm Petrel, *Oceanodroma leucorhoa*, has a wonderfully descriptive name meaning 'white-rumped ocean racer'. William Leach (1790–1836), a British zoologist, was honoured by Vieillot in the naming of the bird. (Here, and in the next two names, the BOU prefers to omit the word 'storm'.)

Swinhoe's Storm Petrel, *Oceanodroma monorhis*, was first described by the English naturalist Robert Swinhoe in 1867. This one is also an 'ocean racer', but with a 'single nostril'.

Wilson's Storm Petrel, *Oceanites oceanicus*, is named for the Scottish-American Alexander Wilson (1766–1813), the 'father of American ornithology'. The Oceanides were mythical sea-nymphs – an appropriate name for one of the most numerous of ocean racers.

The Madeiran Storm Petrel, *Oceanodroma castro*, has a specific name which appears to be derived from the Portuguese for a 'castle', though it seems that Alec Zino felt that it may well be a local word to describe the brooding calls. The British-educated Zino, of course, has his own petrel, *Pterodroma madeira*, a species considered extinct until he discovered a residual breeding colony in the mountains of Madeira in 1969. Zino's is not a 'storm petrel', but rather a 'gadfly petrel', a further subgroup of the Procellariiformes. The fluttering flight is generally considered to be the origin of the term, but Warham drew attention to the high arcing flight which recalled the action of a horse maddened by the bite of the fly. In fact, an old name, *Oestrelata*, meant 'goaded by a gadfly'. That word was misspelt by Bonaparte as *Aestrelata* when he applied it to the Capped Petrel. Zino's was eventually confirmed as a separate species from the similar, but larger, Fea's Petrel, *Pterodroma feae*. Their generic name means 'winged racer' and the latter species records the Italian explorer-naturalist Leonardo Fea (1852–1903). Amazingly, a third petrel is also known to breed in Madeira: Bulwer's Petrel, *Bulweria bulwerii*, was named after an English parson, James Bulwer, who collected a specimen during a trip to Madeira in 1823.

For most of us, our awareness of the Procellariiformes is a matter of the occasional sea-watch, or the rare pelagic boat trip, and the vague chance of connecting with a recognisable dot on the sea. Consequently the names of most of them will remain more familiar than the birds themselves.

WITH MY CROSSBOW... (ALBATROSSES)

And, though I have never yet seen an albatross, the strange evolution of its name is well worth consideration. It seems that the word derives from an Arabic word for 'pitcher' (a large jug). This then evolves in Spanish and Portuguese to become a word for 'bucket', a version of which becomes *alcatraz*, and eventually fixes on the huge scooping bill of a pelican. Other large water birds, such as Gannet, are then included under that umbrella name, with an Anglicised form, *alcatras* (1564), becoming a name for frigatebird. From there, the name further evolved, via the bridging form *algatross*, into the current form by 1681.

THE ANCIENT MARINER

Linnaeus created the genus *Diomedea*, from which followed the family Diomedeidae. In choosing this form, he was recycling a seabird name used by Latin authors, but without any exact reference. The source of this name is an ancient legend which told, in various versions, that the death of King Diomedes so upset his companions that they all turned into seabirds or, alternatively, that the albatrosses gathered to sing in lament. Wandering Albatross, *Diomedea exulans*, was described by Linnaeus in the *Systema Naturae*, and the name reappears more appropriately in that of the Mediterranean Scopoli's Shearwater, *Calonectris diomedea*.

The lone Black-browed Albatross seen frequenting a Gannet colony in Shetland a few years ago became known as Albert Ross, which is a curious modern variation of the process which created names such as Magpie and Robin. At the same time it pays unwitting homage to the maritime traditions which gave rise to names like Tom Harry for the Great Skua and the William link in Guillemot. Albert joins the long-staying Sammy the Stilt, of Titchwell fame, to give two recent examples of a phenomenon which is rooted in time.

The origin of the albatross as a burden is down to Coleridge's poem 'The Rime of the Ancient Mariner', but a little later the French poet Baudelaire also turned it into a powerful metaphor for frustrated genius in 'L'Albatros'. In the former, the sailor who shoots the albatross with his crossbow brings bad luck on the ship and is forced to wear the dead bird round his neck, as in Doré's image here. In the latter, the magnificent bird is grounded on the deck, powerless, and mocked for its clumsiness by the heartless sailors.

So next time you take a pelagic trip, please leave your crossbow at home.

IF ANYONE CAN... (PELICANS)

I find it odd that I have as yet to see the two European pelican species, since I have spent hours watching both Brown Pelicans and American White Pelicans in Florida, and have seen the Australian Pelican from Kakadu to Tasmania.

The prison of Alcatraz in San Francisco harbour got its name from the Brown Pelicans which roosted on the island. However, a Wikipedia article on the origins of the word 'albatross' is wrong to suggest that the Arabic root-word *al-gattas* meant pelican. As quoted by Jobling, Moore is quite clear in stating that *alcatraz*, meaning pelican, is 'itself a corruption of the Arabic *al-gattas* meaning diver or plunger'. That is quite a different matter, especially since no Old World pelican species dives. Instead, they use the quite different technique of creating a feeding line of swimming birds, as do Australian Pelicans. The Old World origin of the word suggests that is was probably first applied to another large seabird, such as the Gannet, while its subsequent evolution into the word 'albatross' underlines the open-endedness of its use among mariners.

BROWN PELICAN

Linguistically speaking, pelicans are pretty plain fare, with the Greek *pelekan, pelekanos* at the root of their name, though that word is derived from *pelekys* (axe), an obvious reference to the bill. The Latin for Pelican is *onocrotalus* (from the Greek *onokrotalos*, meaning 'pelican'), and that joins the Greek as the specific name of the Great White Pelican, *Pelecanus onocrotalus*. The Dalmatian Pelican, *P. crispus*, seems marginally more interesting, with a specific name meaning 'curly-headed' – the 'unkempt look', according to the *Collins Bird Guide*. The American White Pelican, *P. erythrorhynchos* (red-billed), is named for the fact that it gains a rich orange-red bill in the breeding season. The Australian Pelican, *P. conspicillatus*, has a distinctive eye-ring, so is 'spectacled'.

AUSTRALIAN PELICAN

Like the albatross, the pelican is a bird with a long history of symbolism. An older legend, which told of the parent drawing blood from its own breast to feed its young, was adopted by the Christian Church in the second century as a symbol of self-sacrifice. Alfred de Musset, the French Romantic poet, used the idea to great effect to symbolise the role of the poet suffering for his art, a concept echoed later by Baudelaire's aforementioned albatross image. The ornithological truth behind the legend is, of course, the fact that the young thrust their heads deep into the parental pouch to receive regurgitated food, but I suppose that the bleeding heart made better headline material when you wanted to sell a concept. But that does remind me of one memorable individual which sat forlornly in a harbour in New South Wales with a massive fish, too big to swallow, wedged firmly in its throat, apparently waiting for the head end to be digested before the remainder could go down. Now if you want a symbol for gluttony...

TAKING THE PLUNGE: GANNETS AND BOOBIES

As explained earlier, there was not much attempt in early English usage to separate Goose and Gannet, which had common linguistic roots. The latter was long known in Scotland as the Solan Goose. The word *solan* entered the English language via Gaelic, but was derived from the Norse for Gannet, *sùla*, which was to become the generic name *Sula* (as in *Sula bassana*). Some of the gannets, including the original Solan Goose, now the Northern Gannet, have now been moved into the genus *Morus* (from the Greek mōros, foolish, which of course relates to the name 'booby' that is given to some other gannet species). Ironically, the boobies remain in the genus *Sula*. The collection of eggs and young Gannets was once the habit in some cultures, and the fearlessness of the adults in their breeding colonies made them easy prey to humans, hence the second term. The converse of that is that the word 'gannet' is widely used as a derogatory term for someone who is excessively greedy.

With the change of the generic name, the word *bassana* remained in use, but with its gender changed to masculine in *Morus bassanus*. It derives from the Bass Rock in the Firth of Forth, where huge numbers have traditionally nested and which is currently the largest colony of Northern Gannets in the world. Two local names for the bird were Bass Goose and Basser. The word 'gannet' seems to have first been linked in 1274 to the Gannet's Rock (*Petra Ganetorum*), a nesting colony on Lundy in the Bristol Channel. For various reasons, which Lockwood feels might be linked to a sort of taboo relating to the occupation of the island by Norsemen and then by pirates until late into the Middle Ages, the word was not taken into more general use until later, then adopted by Pennant in 1768, and formalised by Yarrell in 1843.

NORTHERN GANNETS

The Gannets which I saw in Quebec and off Prince Edward Island in Canada represent the smaller, western part of the population of Northern Gannet. In winter, the birds range further south and may even overlap, off the coast of Africa, with the very similar Cape Gannet, *Morus capensis.* There is no strong evidence that the populations interbreed. However, there there was a curious mix-up a few years ago, when a British postage stamp depicted a gannet colony. A sharp-eyed bird enthusiast pointed out that these were in fact Cape Gannets. The result was a minor storm of embarrassment for the Post Office.

Even if the terms 'booby' and 'gannet' are unflattering to humans, I would readily bet that huge numbers of birders have imagined, as I have, the sheer exhilaration of being a Gannet and of making those soaring flights and breathtaking plunges. Not only are they acrobatic in the air, they have enormous stamina: those I saw recently off the Needles in June had probably come from across the Channel, while large numbers off the Farne Islands in Northumberland during the breeding season came from the Bass Rock itself. Likewise, birds which I saw off Prince Edward Island in Canada could have flown from Percé in Quebec. Those are not inconsiderable distances to go for a day's feeding.

BALD RAVENS: CORMORANTS AND SHAGS

One of the most intense memories I have of my day on the Outer Farnes is of a pair of Shags, *Phalacrocorax aristotelis,* sitting on a nest-pile of gleaned campion while three of us consumed our picnic lunch perched on a rock not six metres from them (the smell of guano is a curious accompaniment to sandwiches, I recall). The birds were completely comfortable and we had plenty of time to admire their beautiful green gloss and sinuous shapes. The word Shag was first applied to describe the crest on the head of this cormorant and was first noted in 1556. Before that date, and for long afterwards, there was much confusion of this species with the larger, but similar Cormorant. The derivation of the latter word is via French from the Latin *corvus marinus,* meaning sea crow. That in turn is echoed in part in the Greek form of the scientific name, *Phalacrocorax,* meaning 'bald Raven'. Today, of course, both words are used as generic English names for other cormorant species, so these two have become European Shag and Great Cormorant. There is no consistency in the use of the two names with reference to the crests, as the North American Double-crested Cormorant illustrates. This species, *Phalocrocorax auritus,* shares part of its name with the Slavonian (Horned) Grebe. In this case *auritus* is more appropriate, since two ear-like tufts are displayed in the breeding season.

SHAG

The Great Cormorant is widespread throughout the world, though it goes variously by names such as Great Black Cormorant, Large Cormorant and Black Shag. That particular cormorant is indeed much larger than others, a factor which I appreciated most in Australia, where we came across the Little Black Cormorant, *Phalacroroax sulcirostris* (furrow-billed). Australia also had two common pied species, and we found a third, the Black-faced Cormorant, *P. fuscescens* (blackish), in Tasmania.

GREAT CORMORANT

Australia is also home to one of a closely related family, the Australasian Darter, *Anhinga novaehollandiae*, a cousin of the Anhinga of the Americas (of which more later). These birds have a cormorant-like body, but a longer, very sinuous neck, which darts forwards to spear fish prey, hence the common name. An alternative term, the Snake-bird, derives from their habit of swimming with just the neck sticking out of the water.

THE WORLD OF OLD NOG: HERONS, EGRETS AND BITTERNS

I was first introduced to Old Nog the Heron, in Henry Williamson's iconic story *Tarka the Otter*, when I was about ten years old. Williamson had probably made up the name, but it was such a strong idea that I still see Grey Herons as Old Nog, and I daresay that others do too. Williamson's writing was one of the strong influences which drew me towards the natural world and which see me writing this now.

Broadening horizons

A case study of the various words for heron shows why a universal scientific language was both good and necessary, but it also shows that in the post-Linnaean world things are not always that simple.

The Heron (Grey Heron) was known variously in Scottish and English regions by such names as Hegrie (a form related to Egret), Herl, Hern, Heronshaw, Hernsew, Heronsew, Hernsaw, Harnsaw, Handsaw and Heron. If you have spent hours wondering why Hamlet couldn't tell a 'hawk' from a 'handsaw', all is now clear. You can sleep in peace at last: he wasn't confusing falconry with carpentry after all. What is more, Choate points out that those variants of 'shaw' had already confused a word meaning 'wood' or 'thicket' with the diminutive French ending *–ceau*, as in the Old French *herounceau* for a young heron. As so often, Thomas Pennant's preference for a name was influential, and his use of Heron became the norm for later scholars.

The variety of dialect names was almost certainly repeated in many countries of Europe, and even today a small sample of modern vernacular European names shows how easily national languages can become unhelpful: *Grahäger, Blaue Reiger, Graureiher, Héron cendré, Garza real, Airone cenerio Czapla siwa, Szürke gém,* Σταζτοτσικνιας... Few of us would recognise many or any of those names without further help.

GREY HERON

In short, the situation was chaotic before Linnaeus put in place a system by which all living creatures could be classified by scientists in a universal language.

Even once the word 'heron' had been stabilised in English usage, a different problem occurred when we started to look outside our corners of rural Britain, where we normally only saw one heron species. Others showed up overseas, so we eventually needed prefix names like Grey, Purple and Great Blue to distinguish the many different species of heron encountered. Some of these exotic species turn up as vagrants, or more recently as immigrants, to swell the British list. It is also sometimes handy when abroad to check the scientific names to be sure that a local name does in fact depict a different species: I remember that my first Great Blue Heron in Canada was sufficiently similar to Grey Heron to warrant a check of the scientific name: it was *Ardea herodias*, not *Ardea cinerea*. I had already been disappointed to find that the Black-bellied Plover was in fact our Grey Plover, so was glad to see that this one was different. And just in case there are those who think the difference is obvious, it was not so clear through an indifferent pair of binoculars across the vast marsh at Rimouski on the St Lawrence River. I later got to see them much better and could see the difference more clearly.

A plethora of herons and egrets

A further problem started to occur when I went to Florida. There I was faced with a real assortment of 'herons' and 'egrets', some of which were confusingly named. One bird was known locally as Louisiana Heron, but was also called Tricolored Heron, and to make matters worse it was *Egretta tricolor*. Then there was the Great Egret (or Great White Egret) which was sometimes called the Great White Heron (a form which appears in Coward's version of the BOU list in 1926), while, apparently, a race of the Great Blue Heron had proper claim to that last name. You begin to see why I was bemused. Eventually, after several visits, I saw all the available species, colour-morphs and age-plumages of the Florida herons and egrets and became quite confident, but it did take time. I was more prepared when I first tackled the numerous Australian species. But the more I learned about herons and egrets the more complicated the family became, because subspecies crept in too.

In general, the names 'heron' and 'egret' seem to be used quite loosely, particularly since the scientists seem divided about what is classified as *Ardea* and what is considered *Egretta*. A rule of thumb might be that big ones are herons and smaller ones are egrets. That should strictly be true, given the roots and evolution of the two words. In Old French, a word quite similar to heron, *haigron*, seems to be at the root of the word egret too. The loss of the *h* leaves *aigron*, of which the diminutive is *aigrette*, which travels into Medieval English via *aigret* to become *egret*. Voilà!

If it is that simple, why are Great Egrets larger than some herons? Why are Green Herons one of the smallest of all? And why are half of the *Egrettas* known as herons? This confusion is fairly typical of many issues to do with both vernacular and scientific bird names, though in this case there is less division by continent than in the naming of raptors. In short it all boils down to the fact that it doesn't really matter too much in the vernacular forms, even if it is inconsistent, simply because 'heron' and 'egret' are so closely related linguistically.

In terms of scientific classification, however, there are recognised differences within the Ardeidae, though the exact status of some species is hotly disputed.

The Great Egret, in particular, has long been in a state of flux: the one stable factor has been the specific name *alba*, but is the correct generic form *Ardea*, *Egretta* or *Casmerodius*? And for some there is a case for considering whether its four races (American, Eurasian, African and Eastern) are in fact full species. The current position as set out on the IOC list is that it is *Ardea* and that there are four races of one species (though with a note that the American race may yet be split). Great Egret is the IOC form, but Great White Egret appears in the BOU list, so both forms occur in references.

Determining the dividing lines between species is still not an exact science, but experts have certainly been divided about whether the Western Reef Egret/Heron (now *Egretta gularis*) is merely a specialised Little Egret, while the Pacific (Eastern) Reef Heron/Egret, *Egretta sacra*, seems to be unchallenged. Again IOC currently lists them as separate species. Meanwhile, as a layman, I am not in a position to make a definitive judgement. I am even more confused when I realise that Grey Herons have been known to hybridise with Great Egrets. Science itself seems to be in a state of flux: DNA studies and fossil evidence are now challenging past classification patterns in a big way. Meanwhile I will sit and wait for the jury to return and go on enjoying my birds, as well as some of the other linguistic issues which the family reveals.

Behind the names

A fairly obvious place to start is with the word *Ardea* itself, which is a metaphor: it was chosen by the Greeks in recollection of a town of that name which was destroyed by fire. The Grey Heron's specific name *cinerea* (ash-grey) underlines that point.

A trawl through the other large herons which I have seen gives me the following specific names: *herodias* (Great Blue) from a Greek word for heron with a link to a legendary figure, Herodias (see Appendix); *cocoi* (Cocoi or White-necked) from a Cayenne Amerindian name for the local species; and *sumatrana* (Great-billed, a.k.a. Sumatran).

Among the egrets, the Australian Pied Heron is *Egretta picata*, apparently 'daubed with tar', while Little Egret is *E. garzetta*, an Italian name for the species. A Chilean, Molina, was responsible for a linguistic gaffe in 1782, which gave the Snowy Egret the name *thula*. That is in fact the Aruacano Amerindian name for Black-necked Swan. Well, at least is it a water bird and mostly the right colour.

And thinking of colour, my earlier use of the phrase 'colour-morph' reminds me that the Reddish Egret, *Egretta rufescens*, gave me some troubles, simply because it comes in two completely different colours – reddish and pure white. A similar problem occurs with the Little Egret too, though examples of the dark blue-grey morph are virtually unknown in Europe. My real point is, though, that this phenomenon is also sometimes called a colour 'phase', which implies, quite inaccurately, that the bird will grow into another phase (as with the Little Blue Heron, which is white as a juvenile). The colour-morph is not a fault, but merely a genetic element which produces two plumage types within a species – and these are fixed, not transient.

LITTLE EGRET

The broader family

Squacco Heron, *Ardeola ralloides* is currently my bogie bird among the herons. The reason is simple: I have seen all possible members of the family in both North America and Australia (with the exception of two bitterns in each continent), plus a few more in Trinidad, yet this one European species has always evaded me. But I won't hold that against it here, as the name is too odd to miss out. Willughby first noted it around 1672 as *Sguacco*, from an Aldrovandus reference of 1603. It appears that it was probably a purely local Italian dialect word describing the bird's strange voice. Ray used the current spelling and it stuck. *Ardeola* simply separates the species from the mainstream herons and egrets and is a diminutive of *Ardea*. The specific name, *ralloides*, likens the bird to a rail (*Rallus*).

Striated and Green Herons underline the looseness of the terms 'heron' and 'egret', because they belong to the genus *Butoroides* (meaning that they 'resemble bitterns'), which constitutes a further branch of the Ardeidae. That analogy is certainly true of their shape and of the plumage of the young. They differ in that the adult develops a more heron-like plumage. The BOU and IOC lists have both adopted the American split of Green Heron, *B. virescens*, which used to be considered a race of Striated Heron, *B. striata*. We saw the nominate race of the latter in Trinidad and saw two more races in Darwin and New South Wales respectively. In between, there is any number of other races, including one sometimes known as the Mangrove Heron, which is beautifully depicted in the mosaics of the House of Birds in Italica. It seems possible that more splits will eventually occur within this species.

CATTLE EGRET

Cattle Egret, *Bubulcus ibis*, takes its generic name from an idea first used by Audouin (whose name appears in the gull species). *Bubulcus* is the Medieval Latin for a cowherd, so of course fits very nicely for a bird which feeds in the company of livestock. It is of relatively recent usage, because in Coward's 1926 list it appears as *Ardeola ibis* alongside the Squacco Heron. However, *ibis* starts its route towards this name in the mouth of a persuasive Egyptian guide, who convinced Hasselqvist that the bird was the Sacred Ibis. Later, Linnaeus himself perpetuated the error when he created the formal name of the bird. Coward's vernacular version was the Buff-backed Heron, with reference to the buff tones which highlight the bird's plumage in the breeding season. Now it has to be formally called the Western Cattle Egret, since the former subspecies is now the Eastern Cattle Egret, *Bubulcus coromandus*. We saw this species in Australia, though we missed seeing the much more complete buff-apricot plumage which it adopts in the breeding season. The word *coromandus* relates to Coromandel, a word greatly favoured in the eighteenth century to describe part of the coast of India. The American population is the same species as that of Europe and Africa, having first established from vagrant birds which crossed the Atlantic in the nineteenth century.

Night Herons form another secondary group within the Ardeidae. I first met the Yellow-crowned Night Heron and the Black-crowned Night Heron within a few days of each other in Florida, so, unlike the situation at home, I needed the full international names and spelling from the start. The BOU prefers the form Night-heron (with a hyphen). We have only the one in Europe, strictly also the

Black-crowned Night Heron, but of the nominate race. I later saw the Nankeen (or Rufous) Night Heron in Kakadu. In the last case the IOC list has confirmed the older name Nankeen, which refers to the reddish tan-coloured cloth once traded under that name. So in those three species we already have a number of issues in the details of vernacular usage.

As for the scientific forms, the Yellow-crowned is *Nyctanassa violacea*, which translates as the 'violet-coloured queen of the night'. The subspecies which we saw in Trinidad is *cayennensis* (Cayenne is now French Guiana). Those names are a little more glamorous than those of Black-crowned, *Nycticorax nycticorax*, which tell us in stereo that it is a 'night-raven' and also a bird of ill omen in Greek. The American subspecies of the latter is *hoactli*, the Aztec name of the species.

Nankeen Night Heron is *Nycticorax caledonicus*, the specific part of the name being derived from New Caledonia, a group of islands off Australia first discovered by Captain Cook and so named because they reminded him of Scotland (today they are French territory).

EURASIAN BITTERN

Bitterns might be considered to be at the other end of the continuum from the *Ardea*/*Egretta* part of the family. As with the Heron, we can generally refer to the one British species as *the* Bittern, but we also need to allow for the possible vagrant American Bittern and Little Bittern, so it has now become the Eurasian or Great

Bittern, *Botaurus stellaris*. The word Bittern has its roots in Latin, since *butire* is the verb 'to boom'. Its nickname *taurus* (the bull) became attached too, and it entered English as the Medieval French form *butor*. From that it travelled through *bitore* to *bitter* (with a few variants on the way) and then to the modern spelling, which was standardised by Pennant in 1768. *Botaurus* returns only to Medieval Latin for its source, while the word *stellaris* describes the black starry patterns of its upper plumage. For British birders, it is worth bearing that last name in mind, since the American Bittern lacks that characteristic to its plumage, having a relatively plain mantle: one just may turn up on your patch.

Little Bittern is *Ixobrychus minutus*, so the scientific name warns us that the two bitterns are not so closely related, although the plumage of the juvenile bird is a pretty good miniature version, as I found when I was lucky enough to see both a Great Bittern and a juvenile Little Bittern within minutes of each other at Titchwell in 2011. *Ixobrychus* derives from the Greek words for a reed species and the equivalent of *butire*, while *minutus* is no exaggeration: it is smaller than a Coot.

I was lucky enough to see the Black Bittern, *Dupetor flavicollis*, in Kakadu, at the wonderfully named Barramundi Creek. *Dupetor* apparently means 'clatterer', but since there is no record of bill-clattering this name seems to be inaccurate. It certainly does not relate to the voice, which is described as 'a low booming drum-like call'. Its black plumage is perfectly adapted to a life in the deep shadows under palms and *Pandanus* scrub. As *flavicollis* suggests, there is some yellowish plumage at the neck, which may just serve to break up the silhouette.

OF STICKS, SPEARS AND CHILDREN: STORKS

The White Stork, *Ciconia ciconia*, is only an unfamiliar vagrant to Britain and Ireland, but it holds a major place in the culture of mainland Europe. It is in fact rather hard to ignore a bird which places a pile of sticks and mud up to two metres high on the roof of your home, church or town hall.

WHITE STORK

The scientific name is quite easy, because it is the Latin for the White Stork. As for the word 'stork', that throws in a few more complicated elements. The OED suggests that it is of Germanic origins and relates to the verb 'to stalk', the implication being that the stiff posture and stalking habit has led to the name. Lockwood affirms the link to the Germanic *storch*, which can also mean 'stick' (stalk in another sense), and which could link to the nest. However, he points out that the word also means penis, 'this evidently being the source of the (hitherto unexplained) fable that the Stork brings the babies'. There is a curiously related link to that story in that the Binbinga people of Australia have a taboo against the eating of the meat of the Jabiru Stork, fearing that to do so would cause the unborn child to kill its mother. According to Gould it is foul and fishy anyhow. A further superstition from Estonia links the stork with death, but this may relate to the Black Stork, *Ciconia nigra*, rather than the White.

From the above, it is clear that the word 'stork' has become a widespread generic name. The storks are represented by a number of genera and species worldwide, one of which is the statuesque Jabiru of Australia. Though widely used locally, that word is in fact imported from the name of a broadly similar South American species, *Jabiru mycteria*. The original is a Tupi Amerindian word, which was adopted into Brazilian Portuguese and became widely known among explorers and travellers. It has a certain attraction and colour as a name, but the duller truth is that we should call the spectacular Australian bird, *Ephippiorhynchus asiaticus*, the Black-necked Stork. Having seen the bird in the sunshine of Kakadu, I can be sure that the word 'black' is a most inadequate way to describe the iridescent dark turquoise which dominates the head and neck. As for that strange Greek word *Ephippiorhynchus*, we have Temminck to thank for that: *ephippio* means 'saddle' and *rhynchos* is 'bill'. A related species, Saddle-billed Stork, *E. senegalensis*, has a dramatic red patch over the bill to justify that name. The feature is absent in *asiaticus*, which name implies that it is also an Asian species. The Australian bird's dramatic all-black bill gave rise to one Aboriginal legend that the bird's head had been pierced by a spear.

I had previously got to know the Wood Stork, *Mycteria americana*, in Florida. As its name suggests, it belongs to a third branch of the family. *Mycteria* derives from the Greek *mukter* (snout), the bill being once again the outstanding feature (though not as dramatic as in the previous species). One of my favourite recollections is of an encounter in Florida during an early-morning stroll. We passed an elderly couple taking breakfast on the lawn: their three live 'garden ornaments' were a Wood Stork, a Snowy Egret and a Great Egret, which were, they told us, regular guests. So much for the Wood Stork's rarity! There were not many in the area, but they were frequent enough for us to see them several times during each visit.

As for Black Stork, our one encounter with those was dramatic, but more distant, as a flock of thirty – a significant percentage of the Spanish population – struggled to make landfall near Tarifa in a very stiff easterly wind. White Storks were frequent in Andalucia, on buildings and in the fields, but I had known those for many years.

VENERATION AND SPATULATION: IBISES AND SPOONBILLS

My spell-checker tells me that the word 'spatulation' doesn't exist. Well it does, but not necessarily in the way intended. I have bent the tool to my own use. I once learned how to turn old screwdrivers into ad hoc carving tools, and I also hint later on that I might not be averse to using a knife as a screwdriver, so this is the linguistic equivalent. Newly coined words are neologisms. The technical term for this ploy is 'cheating'.

This section is about birds of the family Threskiornithidae, which means 'worshipped birds'. The Sacred Ibis is a clue (unlike the Scared Ibis, which is a typo). The close relatives of the ibises are the spoonbills (hence the spatula).

We are now starting to see Glossy Ibis and Eurasian Spoonbill in increasing numbers in Britain. The former species now seems to appear regularly after the breeding season in Iberia, and I have seen it here on seven different occasions in recent years. The Spoonbill now appears annually on my British list, with a 2012 sighting on the south coast of seven birds in flight. It has already bred in Norfolk.

The word 'ibis' is of Greek origin, though the sacredness of the one species was down to the Egyptians. This may well have grown from the birds' propensity to hang around human habitation and, like the cat, to help rid homes of vermin. Glossy Ibis, *Plegadis falcinellus*, obviously takes its vernacular name from the gorgeous iridescent plumage. *Plegadis* is from a Greek word for 'sickle', while *falcinellus* derives from the Latin for the same idea. This root occurs also in the origin of the word 'falcon', though in the latter case it refers to the shape of the wings rather than of the bill. I have also met Glossy Ibis in Florida, where the American White Ibis, *Eudocimus albus*, was more common in the area which I knew best. One of my great pleasures was to watch their elegant dawn flight over Tampa Bay. An evening session to watch the Scarlet Ibis, *Eudocimus ruber*, roost in Caroni Swamp, Trinidad, was like watching poppies driven by the wind to festoon the mangroves. If I had been naming the species, *papaver* would have been in there somehow. *Eudocimus* means glorious, so that is quite apt. The Australian White Ibis, *Threskiornis moluccus*, has a black head and little pride, since I have seen it scrounging round urban picnic areas and one of my friends actually photographed one inside a rubbish skip. I liked it a lot better in open swamps, where it was alongside more Glossy Ibises. The specific name relates to the Moluccas (modern Maluku), an island group now within Indonesia. The Straw-necked Ibis, *Threskiornis spinicollis*, was impressive on an NSW golf course, but I was almost assaulted by one in Kakadu when it was my turn to guard the picnic lunch while the others watched rarer birds. The bird gets its vernacular name from the shaggy yellowish plumage on the throat, which also leads to the less satisfactory scientific name, meaning 'spine-collared'.

WHITE IBIS

In Andalucia we had no time to look up the reintroduced Bald Ibis, *Geronticus eremita*. With only an estimated 400 birds in the wild, the 2013 estimated population of the Barbate project was an invaluable eighty birds. *Geronticus eremita* means 'old man of the desert' (rather than 'old hermit', which of course is linked linguistically).

Eurasian Spoonbill, *Platalea leucorodia*, has a name derived from the spatulate shape of the bill. Charleton first used the term Spoon-billed Heron in 1668, but ten years later Ray used the form 'Spoonbill', after a Dutch example. Pennant and others followed this form. The Greek work *platea* (broad) is at the root of the generic name, while *leucorodia* derives its form from *leukos* (white) and *erodias* (heron). The Royal Spoonbill, *Platalea regia*, which I saw in Australia, is more closely related than the Roseate Spoonbill, *P. ajaja*, of Florida. In that case the Tupi Amerindian word for the species, *ayaya* or *ajaya*, is the source of the scientific name. Roseate is a pink as pink could ever be, when set off against a cobalt sky, as we witnessed when one flew low over our heads while we watched from a jetty on Tampa Bay. When Mary asked why I had not raised the camera, I had to explain that I was quite overawed and didn't wish to spoil the moment. I have never regretted that decision, since the mental image is so strong.

EURASIAN SPOONBILL

ROSY FACES, ROSY BIRDS: FLAMINGOS

One of the great delights of our trip to Andalucia was to see the large flocks of flamingos at Fuente de Piedra and elsewhere. These were, of course, Greater Flamingo, *Phoenicopterus roseus*.

GREATER FLAMINGOS

The story of the name 'flamingo' begins with rosy-cheeked, blond, medieval Flemish sailors (Flemings) appearing as traders among the sallow-skinned inhabitants of Lisbon. From that, the locals gave the rosy bird the nickname *flamingo* and it stuck. It is ironic that the Spanish dance, the Flamenco, owes its origins to the flamingos directly, but indirectly to the Belgians. Today, just a single letter differentiates in French between the people and the birds: *flamands* and *flamants* – which just might keep the Brussels translators on *their* toes. A document headed 'La conservation du flamand rose' would really square the circle.

That wonderful generic name *Phoenicopterus* is the Latin version of a Greek word meaning crimson-feathered. This is a surprise choice to anyone who has not seen a Greater Flamingo in flight, since the bird on the ground simply lives up to the specific name *roseus* (pink). It is the flight feathers that reveal the dramatic crimson. Chilean Flamingo and American Flamingo share that generic name, but very recently Lesser Flamingo has been given the unique generic name *Phoeniconaias* (a scarlet naiad), while Andean Flamingo and James's Flamingo share the genus *Phoenicoparrus*, meaning 'a scarlet bird of omen'.

While the names Greater Flamingo and Lesser Flamingo look illogical in the context of all six species, they did originally separate just the two Old World species.

A TANGLE OF CROOKED BEAKS AND CLAWS

The Willughby/Ray classification of 1678 divided 'land birds, with crooked beaks and claws' into 'diurnal birds of prey', 'nocturnal birds of prey' and 'frugivores'. The detailed examples were dubious, but clearly there was an attempt to recognise common characteristics. Today we still use the term 'birds of prey' or often 'raptors' for the hawks, eagles and others in the diurnal group, but would exclude shrike, which was on the Willughby/Ray list, because we now consider them to be passerines.

A further complication is that we now tend to exclude owls from the umbrella term 'birds of prey' or 'raptors' (even though both terms fit their behaviour). Because the species, like many owls, was nocturnal, Willughby and Ray included the 'goatsucker' (nightjar) in the same category as owls, but that may simply have been more for its alternative folk-name, Fern Owl, than for any other feature. As for frugivores, or fruit-eaters, these were parrots and cockatoos, which today stand out like a sore thumb in the company of the hunting species. In short, time has changed both knowledge and fashion. We now know that raptors, owls, nightjars and shrikes belong in four distinct and separate categories. We also now know that the hooked beak and claws of three of the groups demonstrate convergent evolution in which birds of prey, owls and shrikes arrive at some similar characteristics for the similar jobs of catching live birds, mammals or reptiles, and for holding and tearing their prey. In the case of the frugivores the solution was similar but for dissimilar reasons: instead of needing to kill, and tear flesh, these birds needed to lever, crack and hold. In need, a butcher's cleaver will hew wood, or a dinner-knife drive a screw.

Owls are indeed birds which prey on other creatures, but it is not facetious to point out that such a definition can apply to the vast majority of birds. Even Chaucer noted this of the swallow:

> *The swalwe, mortherere of the foules smale*
> *That maken hony of floures freshe of hewe*
>
> (The swallow, murderer of small creatures which make honey from the fresh-coloured flowers)

We would probably term the Heron or the Great Grey Shrike as 'predators', but not think of using that term for the Mallard, Oystercatcher, Nightjar or Song Thrush, which all prey on other creatures, whether frog, mussel, moth or snail. Even the Blue Tit is a predator of caterpillars. In short, such terminology is values-based, which is why we have today pinned a limiting value on the terms 'birds of prey' and 'raptor' – and the raptors are what I want to look at next.

RANK AND STATUS IN THE REALMS OF RAPTORS

The buzzards are circling

Long before I knew of Buzzards in England, Saturday cinema had introduced me to dying cowboy heroes, suffering and parched in the Arizona desert with the buzzards circling. This turned out to be Western-speak for vultures, of course. Except that they were not true vultures either, in the Old-World sense of the word. American vultures, the Cathartidae, were long thought to be more closely related to the storks, but recent studies of fossil and DNA evidence now show them to be much closer to the other raptor families. There are in fact two fairly widespread species in North America, which I got to know quite well, and both were plentiful in Trinidad.

Turkey Vulture, *Cathartes aura*, has a number of subspecies, even within North America, and some are more migratory than others. Suffice it to say that I have seen individuals in summer as far north as New Brunswick and Montreal – the latter right over the city centre – and as far south as Trinidad, and once watched a spectacular northward movement of them from a Florida jetty. The word Turkey simply refers to a superficial resemblance of the red-faced bird on the ground to a Wild Turkey. I certainly wouldn't fancy one of the vultures, roasted or otherwise, having seen some of the road-kill on their menu.

The scientific name is more useful, in that *Cathartes* means 'cleansing' (as in the word catharsis, used to describe the purging of our emotions by, for example, watching theatrical tragedy). Having negotiated the occasional road hazard of flocks of vultures on Florida road-kill, I can attest that this is a very visible function, and I even saw one come down to the shore to eat up the bones of a fish left by a Bald Eagle. And anything which will clear up a road-kill skunk has to be useful. In the light of that, it might be tempting to think that the bird is worth its weight in gold, but that has nothing to do with the *aura* part of its name, which in fact derives from *Arouá*, a native Mexican name for the species. A recent find of a jade depiction of a vulture in an ancient Mayan tomb underlines the fact that the vulture was a symbol of power in that society.

Still in the New World, the Black Vulture, *Coragyps atratus*, enters the arena with a common name which is somewhat controversial. It has long been known as Black Vulture, but *Aegypius monachus* goes by the same name in the Old World and has done so for much longer. The IOC list and Birdlife International currently seem to favour the use of Black Vulture and Cinereous Vulture respectively, presumably because the Eurasian bird is only black-*ish*. Some authorities prefer the use of American Black Vulture and Eurasian Black Vulture, while others have preferred Monk Vulture for the latter, in line with the specific form *monachus*. So what's in a name? I suspect that traditions will die hard and that there will yet be a fair amount of politicking before the matter is settled.

AMERICAN BLACK VULTURE

As for the scientific name, the first part, *Coragyps*, contains the Greek *korax* (raven) and *gups* (vulture), which clearly reflects the raven-black plumage, and is repeated in *atratus* (black). Those in Trinidad were of the smaller race, *brasiliensis*, but the Florida birds were of the nominate race.

But one name which I will always associate with the Black Vulture is Elvis. In simple terms, Elvis was a rescued orphan which had imprinted on people so intensely that he couldn't be released back into the wild. In any case he thought he was a person and, like a small, excited child, showed off dreadfully to anyone who went to see him. He lived in a rescue unit at Moccasin Lakes at Clearwater, Florida, alongside a couple of injured Bald Eagles and a Red-tailed Hawk. Mary and I would have adopted him at the drop of a hat. We had never seen such humour, personality and intelligence in the behaviour of any bird. But the importance of Elvis was that he was an ambassador for the whole of Vulturedom, an amazing educational resource. Many people took away the message that those 'ugly' black birds which they saw scavenging on road-kill were a lot better than OK, and should be respected. It is natural to be revolted by the sight of their feeding-style, but it is important to understand their role.

The Old World Vultures have a name which has changed little in its evolution from the Latin *vulturius*. It was much later in my birdwatching experience that I eventually saw my first members of this group – and the huge Griffon Vulture, *Gyps fulvus*, was hard to miss. The common name is more interesting than the scientific name, which simply tells us that it is a tawny-coloured vulture. Griffon (from the Latin *gryphus*) is an alternative spelling of *griffin* and *gryphon*, and refers to a legendary beast which the Greeks believed to be the guardian of Scythian gold. It had the head and wings of an eagle and the body of a lion. It seems probable that the beast and the vulture met by linguistic accident, because, as Jobling tells us, in 1544 Turner complained that 'Quite wrongly certain scholars call the Vulture Gryps... the Gryps is a Gryphon.' In short, there was confusion between *gyps* and *gryps*. Undoubtedly the fact that the word travelled via Latin and then Medieval French did not help, since *griffes* are claws in French, and both creatures possess those. Subtle confusions do happen elsewhere in bird names.

GRIFFON VULTURE

My one other Eurasian vulture experience was a sighting in Spain of the Egyptian Vulture, *Neophron percnopteros*, and that too leads us back into Greek mythology, since Neophron was a vengeful deceiver who was converted by the gods into a vulture. The second part of the name derives from the Greek *perknos* (dark) and *pteros* (winged). Neophron's vengeance was wreaked upon Aegypius, whose name has already appeared above as the genus of the Eurasian Black or Cinereous Vulture.

I have yet to see a Lammergeier, whose name is originally German for 'lamb vulture'. It was first used as an English name in 1817, but in the form Lammergeyer. For various reasons Bearded Vulture is now the preferred name on the IOC list. The scientific name, *Gypaetus barbatus*, boils down to 'bearded eagle-like vulture'.

I'll grind your bones...

The strange story of the Griffon Vulture is reflected in the name of the Osprey, which arrives at its current form via an inaccurate transcription in Medieval French, when *osFraie* became *osPraie* (an early example of typographical evolution, or to be more precise of calligraphic evolution). However, an error of a far more fundamental sort had already created *osfraie* as a shortened form of *Ossifraga*, the Latin for 'a bone breaker'. This is so clearly unlike the behaviour of an exclusively fish-eating raptor that the most obvious explanation is an early confusion between two large raptor species. A prime candidate would be the Lammergeier, whose speciality is to break and eat bones which other vultures cannot consume. On the other hand the Egyptian Vulture is known to break Ostrich eggs and animal bones, sometimes by dropping large stones as tools in the process. In terms of size and coloration, the latter is much the likelier candidate for confusion with the Osprey.

OSPREY

And before I leave that story, legend has it that the Ancient Greek playwright Aeschylus was killed when a tortoise fell on his head. The Lammergeier does in fact drop tortoises to break their shells to get at the meat, and both it and Egyptian Vulture are found in Greece, so the story may well be more than legend.

The Osprey is *Pandion haliaetus,* which tells another confused tale. There were two Greek kings call Pandion, but it was actually Nisus, the son of Pandion II, who was transformed into a hawk. The same Nisus pops up again in the name of the Sparrowhawk, *Accipiter nisus,* but some versions of the legend link Nisus to a sea eagle or an Osprey. Scylla, the daughter who betrayed him to Minos, was transformed into a lark, seabird or heron destined to flee in fear from her angry father. Incidentally, the two daughters of Pandion I were Philomela and Procne, one of whom turned into a Nightingale and the other into a Swallow, linguistically, as well as physically. The French zoologist Savigny seems to have remembered only part of the story when, in 1809, he created the new name for this genus by confusing the father and son of the legend. It is interesting, though, that 'avian transmutation' was clearly a professional hazard among royalty at that time.

As for *haliaetus,* it derives from the Greek *haliaietos* (sea eagle/Osprey), as does the confusingly similar word *Haliaeetus,* the modern genus of the sea eagles. So why, I ask myself, do we have to wade through so many complicated combinations of vowels?

And before I leave the linguistic tangle of the Osprey, I used to be able to say that it was one of just two species, with Ruddy Turnstone, that I had seen in every country where I have watched birds. They were so common in Florida that I once responded to the question 'Much about?' with the answer 'Only five Ospreys and two Bald Eagles.' We have also frequently seen them in summer in the maritime states of Canada, though Trinidad produced only one sighting in a week there. We saw one in Andalucia, and at home I now have two or three sightings a year. But things have now changed, because, according to the current IOC list I have an unexpected armchair tick: there is now a second Osprey, which is known as Eastern Osprey, *Pandion cristatus* (crested). We saw that in both Queensland and the Northern Territory, when it was still considered a subspecies. The split has left the original species to be known as the Western Osprey.

Fishy tales of ernes and honour

And so back to *Haliaeetus...*

My first concern with this group is whether we should call them 'sea eagles' or 'fish eagles' or just plain 'eagles'. In *Raptors of the World*, Ferguson-Lees insists that they are all fish eagles, which is logical, since not all of them fish at sea. Certainly the option of plain 'eagle' is misleading, since this group is not that closely related to the 'true' *Aquila* eagles. Yet we use White-tailed Eagle and Bald Eagle quite regularly, while sticking to African Fish Eagle and White-bellied Sea Eagle. It is interesting to see those four forms are now the accepted IOC names. The Americans have never had the slightest hesitation about their Bald Eagle, which name separates it from Golden Eagle very nicely. Bald Eagle, *Haliaeetus leucocephalus*, might be considered slightly off-key by those who are losing their thatch, but as a baldy myself, I rather like to be allied to the 'noble' bird. But isn't it ironic that the Americans, in attempting to emulate the Romans, chose not an aquiline eagle, but a scrounging fish-eater as their national symbol? I have watched many a Bald Eagle, and they are pretty impressive in general, but they are not above ambush and highway robbery: I have seen more than one rob an Osprey with skua-like panache. In fact, Benjamin Franklin had the same reservations when he wrote:

> *I wish that the bald eagle had not been chosen as the representative of our country, he is a bird of bad moral character, he does not get his living honestly, you may have seen him perched on some dead tree, where, too lazy to fish for himself, he watches the labor of the fishing-hawk, and when that diligent bird has at length taken a fish, and is bearing it to its nest for the support of his mate and young ones, the bald eagle pursues him and takes it from him... Besides he is a rank coward; the little kingbird, not bigger than a sparrow attacks him boldly and drives him out of the district. He is therefore by no means a proper emblem for the brave and honest... of America... For a truth, the turkey is in comparison a much more respectable bird, and withal a true original native of America... a bird of courage, and would not hesitate to attack a grenadier of the British guards, who should presume to invade his farmyard with a red coat on.*

BALD EAGLE

So there you have it. If nothing else, it is a wonderful example of a very common tendency to condemn wild species for their moral proclivities, as if they might be improved by a good sermon. And it is interesting here to note the once widespread use of the term 'fishing-hawk' for Osprey. A curious fact, which I noted in my own diaries from Florida, is that I sometimes saw Bald Eagles chased off by Fish Crows and Mockingbirds, but never did I see an Osprey attacked in the same way. This could well relate to the fact that the Osprey sticks very strictly to fishing and therefore poses no more threat to other birds than does a tern. The brave kingbird reappears in a later chapter.

Australia also produced a number of sightings of White-bellied Sea Eagle, *Haliaeetus leucogaster*. As much as I admired them, they do little for this narrative, because *leucogaster* simply means 'white-bellied'. But a word of caution – while the adults are slate-grey and white, immature birds are a rather conventional mottled brown, so do not live up to their name. They are still very handsome, though: we saw two adult pairs on the New South Wales coast and a number of varying ages in Kakadu.

As for our own White-tailed Eagle, *Haliaeetus albicilla*, I have seen just the one and an immature bird at that. It didn't have the white tail of either its vernacular or specific name, since that only comes with maturity. This bird had spent much of the winter in the New Forest and then holed up near Basingstoke for a spell in early 2011. Mary and I went down to see it on the very morning that it moved off, and we watched it disappear northwards – roughly in the direction of Earley, where we live.

Now the significance of that may not be plain to all, but the name Earley derives from the erne's lea, or eagle's clearing, *erne* being an Old English word for an eagle. Nearby Woodley also attests to this area being a western part of the great forest that once surrounded Windsor. Today's suburban sprawl would hardly convince anyone of that. Yalden and Albarella summarised work by Gelling and Cole and others, which listed thirty-three place names derived from *erne*. Their summary of historical records also notes that the species is 'well-recorded, much more widespread and numerous than Golden Eagle.' A 2012 study by Richard Evans *et al.* linked Golden and White-tailed Eagles to place names in Britain to help map their former ranges. The outcome is that Golden Eagle occurs in primarily upland names and White-tailed in lowland regions, but vastly more numerous. Archaeological evidence (a bone from a site near Earley and a claw from Dorney, Buckinghamshire), shows that the Thames Valley once held a population of White-tailed Eagles which fished in large natural lakes and ponds.

In short, when our White-tailed Eagle soared away towards Earley, I saw it as something of a homecoming. And the spectacle of the huge bird with no fewer than nine Common Buzzards and two Red Kites 'in attendance' was one of the highlights of my birding life.

The 'royal' eagles

Traditionally, the *Aquila* eagles are seen as the truly regal birds of Eagledom. It is only quite recently that DNA studies have placed Bonelli's Eagle, *Aquila fasciata*, in this category, promoting it from the mere nobility of the hawk-eagles. Professor Franco Bonelli, of Turin University, would have been proud. One showed briefly for us in 2008, but in 2011 we spent a relatively fruitless hour or two watching a known nesting cliff in Andalucia – if you can consider a herd of Spanish Ibex a fruitless session.

From Roman times the great eagles were chosen to signify the status and glory of Empires. For the Romans, the eagle became the symbol which led the legions, one of which features strongly in Rosemary Sutcliffe's *Eagle of the Ninth* (the small bronze eagle which gave her the idea was found at Silchester and resides in Reading Museum). Later the Austro-Hungarians, the Napoleonic French, the Russians and the Germans were to imitate the idea, some with double-headed eagles. The Imperial Eagle was named for this tradition. What is now known as the Eastern Imperial Eagle, *Aquila heliaca,* reflects that importance in a specific name associated with the Sun itself.

This was the tradition which made Franklin feel humiliated by his own country's choice of a 'second-ranker'. His concern about the cowardly eagle would not have occurred in Australia, where Wedge-tailed Eagle is *Aquila audax,* meaning 'the bold eagle'. With an eagle like that, what did the Aussies do? They put a bounty on the eagle, which was a threat to their sheep, and chose the Emu, *Dromaius novaehollandiae,* to partner the kangaroo as their national symbol, presumably on the grounds that the Emu was at least sheep-friendly. Anyone who has ever looked an Emu in the eye will realise that the choice is proof of the famed Australian sense of humour (though I suspect that the Emu was chosen because it is the continent's largest bird). My one close encounter with two wild Emus was unnerving, to say the least. I did at least come to appreciate that *Dromaius* means 'running at full speed': I have never seen wild birds move so fast *towards* me. They had spotted an imminent picnic from a kilometre away and the two small dots grew huge at an alarming rate. Our cowardly retreat would not have impressed Franklin, though Michael Parkinson would have understood.

Fortunately the 'Wedgie', as it is often called by the locals, was granted protection in time. The sight of a family of four of these huge birds circling over farmland north of Canberra was my eventual great pleasure, and a close-up view of a nest in a huge eucalypt on a rocky outcrop of a New South Wales farm was a huge privilege.

My one ever sighting of a Golden Eagle, *Aquila chrysaetos,* was in Switzerland and so far off as to be very unsatisfying. Its specific name contains two elements based on the Greek, *khrusos* (gold) and *aetos* (eagle). I did a little better in Spain in 2008, with a much closer view of the Spanish Imperial Eagle, *Aquila adalberti,* which was truly impressive. The species was named by Brehm after Prince Heinrich Wilhelm Adalbert of Prussia (1811–1873), undoubtedly in acknowledgement of patronage. There was long hesitation before this was regarded as a full species, quite separate from the Eastern Imperial Eagle, *Aquila heliaca.* Its population is small and very vulnerable. Three other species of *Aquila* eagles breed in eastern Europe. Lesser Spotted Eagle, *A. pomarina,* takes its name from the old Baltic dukedom of Pomerania. Greater Spotted Eagle is *A. clanga,* which derives from the Greek *klangos* (eagle). Steppe Eagle, *A. nipalensis,* indicates its more easterly bias, with a name that reflects both the open plains of Asia and the Himalayan kingdom of Nepal.

The 'noble' eagles

If the *Aquila* eagles are the royalty, the nobility is represented by several other groups.

The first are known as the hawk-eagles. I saw just one Little Eagle, *Hieraaetus morphnoides*, in Australia. It belongs to the same genus as the Booted Eagle, *Hieraaetus pennatus*, which we saw in Spain. *Hieraaetus* combines Greek words for hawk and eagle, while the specific name of Little Eagle (rather pointlessly) combines the Greek *morphnos* (eagle) and *–oides* (resembling). The elements of Booted and *pennatus* in the name of the second bird refer to its feathered legs.

In the name of the Ornate Hawk-Eagle, *Spizaetus ornatus*, which we saw in Trinidad, *Spizaetus* says much the same as the previous name, from the Greek *spizias* (hawk) and *aetos* (eagle), while *ornatus* refers to the crest which places this bird in the crested eagles subgroup.

A further group is known as the serpent eagles, though it seems that these are not that closely related to the preceding groups, instead fitting somewhere between vultures and harriers. Another of our Spanish sightings is widely known as the Short-toed Eagle, but is now more formally the Short-toed Snake Eagle, *Circaetus gallicus*. The generic name, like the two previous names, is yet another hybrid form: in this case it is the Greek *kirkos* (hawk) and *aetos* (eagle). The bird is widespread in France, which accounts for the specific name *gallicus* (from Gaul).

What emerges from all of this is that the word 'eagle' is as imprecise as the word 'hawk'. Strictly speaking it belongs to the *Aquila* eagles, since its route into English is via the mutation from Latin *aquila* to Old French *aigle*, which becomes *egle* then *eagle* once it transfers to English. Clearly, both British eagles, Golden and White-tailed, are huge, and it was relatively late historically that naturalists distinguished between the two types which they represented. The word eagle was, and still is, used for other large species.

A second feature of note is that there seems to have been a wide range of different words for birds of prey in Ancient Greek, which implies that there was both a wide vernacular and a great deal of awareness of wildlife in a largely pastoral environment in which there were, undoubtedly, a lot of different raptor species.

Flying a kite

RED KITE

The Red Kite, *Milvus milvus*, has in recent years become a commonplace sight over our home, with often ten birds in the area at once and a record flock of more than thirty over the garden one day in 2015. The reintroduction into the Chilterns in the 1990s was a massive success and the birds have spread out from there. The word 'kite' derives from an Anglo-Saxon form, *cŷta*, an imitation of the bird's whistling call, which can now be heard on a daily basis. The word for the flying toy came after the bird. *Milvus* is simply the Latin for kite. An old name of Norse derivation, Glead/Glede (Gled in Scotland), which Gilbert White related to a Saxon word meaning 'to glide', was used mainly in the in the north of England, but died out during the nineteenth century as the birds became extinct there during the Victorian onslaught on hook-billed birds.

I find that the beautiful bird attracts the attention and interest of non-birders in a positive way, though there is also evidence of some nervousness about hook-billed birds which requires a constant input of facts to offset silly rumours. When a newspaper reported an attack by a Red Kite on a loose dog, it turned out that the pooch was a big as my shoe and unharmed, but the negative effect was sufficient for a local Neighbourhood Watch to circulate dire warnings about pet safety. Mind you, it would never do to circulate the story about the Black Kites of Northern Australia, which have been seen to transfer glowing brands to extend the blaze of a bush-fire in order to flush out more prey items. Imagine the headlines then: *Arsonist Kite Barbecues Chihuahua!*

In 2008 I saw my first Black Kites, *Milvus migrans*, in Spain and then returned in 2011 to watch them live up to their name as they migrated back across the Straits of Gibraltar in a stiff easterly wind. In 2008 I also saw the Australian subspecies (*affinis*) in and around Darwin. As the trinomial suggests this is 'related or similar to' the nominate race: it is in fact a smaller and slimmer bird.

The species which I desperately hoped to see there was the Brahminy Kite, *Haliastur indus*, and was lucky to see it on two different occasions. Appropriately it was over the coast at Darwin, since *Haliastur* is the Greek for 'sea-hawk'. The nominate race is an Asian species, hence both the word *indus* and the vernacular name, which is thought to originate from the white cloth worn over the head by Brahmin priests. While nominate birds have black shaft-streaks in the white head, the Australian race, *girrenera* (an Aboriginal name for it), has a pristine white head and tail which are offset by rich rufous plumage. It is a beauty.

Its congener, the more plentiful Whistling Kite, *Haliastur sphenurus*, occurred inland and was the commonest species in Kakadu. *Sphenurus* means 'wedge-tailed'. When we saw these flying with Black Kites, that feature was important, though the plumage was a few shades lighter too.

We saw other kites in Trinidad, where my favourite was Plumbeous Kite, *Ictinia plumbea*. A bird named after a dull, grey metal, lead, has to be a bit special in shape and style for me to consider it thus.

The most fascinating name surely goes to Trinidad's Double-toothed Kite, *Harpagus bidentatus*, which we sometimes saw perched in the forest canopy below the Asa Wright Centre. Two notches in the bill explain the common and specific names, but the generic name *Harpagus* implies rapaciousness and comes from the mythical Greek monsters the Harpies (as does the name of the Harpy Eagle, which I would love to see!)

When is a kite not a kite?

The word 'kite' is another relatively loose term, since it covers several families of vaguely related species which occur around the world. The *Elanus* kites (another Greek word for kite) fall into this category and include Europe's Black-winged Kite, *Elanus caeruleus*. Past confusion has now been resolved in this genus, with the firm separation of the generally similar species of the Americas, Europe and Australia. The word *caeruleus* implies that the grey of this bird is a bluish colour, but the English name is misleading, since much of the wing is not black. The very similar Black-shouldered Kite, *Elanus auxillaris,* is found in Australia. Its vernacular name derives from the limited patch of black visible on the 'shoulder' of the perched bird. I saw two of these in New South Wales and was able to distinguish them from the Letter-winged Kite, *E. scriptus,* by the fact that they did not have underwing black on the inner wing. Strangely, *axilaris* refers to the featureless 'armpit'. I later realised that the hovering I witnessed was almost sufficient in itself to distinguish between the two.

PLUMBEOUS KITE

We saw Grey-headed Kite, *Leptodon cayanensis,* in Trinidad. *Leptodon* means 'fine-toothed', which I assume refers to its slender bill. But by far the smallest bearer of the name 'kite' was also one of our Trinidad treasures. Pearl Kite, *Gampsonyx swainsonii,* is the second smallest raptor in the world, at about 21 centimetres in length, the size of a Starling. The interesting-looking word *Gampsonyx* is in fact about the most unimaginative name possible for a raptor, another Friday afternoon job: it means 'curved claw'. William Swainson (1789–1855) was an English naturalist whose name also appears in Swainson's Hawk.

My choice of the largest 'kite' might arguably be disqualified, because it is neither true buzzard nor true kite, but Black-breasted Buzzard is also known as Black-breasted Kite. A pair of these dramatic birds was nesting by the trail leading to our camp in Kakadu, and we saw only too well why its unique generic name, *Hamirostra,* emphasises the great hooked bill, which is proportionately worthy of any eagle. The species name, *melanosternon,* repeats the meaning of the common name.

Circuses, ring-tails and swamps

Circus is the generic name for all the harriers and derives from the Greek name of a part-mythical hawk which flies in circles (cf. *Gyro* in Gyrfalcon – and the dubious value of it).

Unfortunately for the Hen Harrier, *Circus cyaneus*, it was accused of harrying hens and can't live down that name and reputation among moor owners and gamekeepers, who still see it as a massive threat to grouse stocks. The opposite case is that the presence of such a raptor species reduces weak stock and genetically strengthens the prey population. Be that as it may, illegal persecution ensures that this species is the most endangered breeding raptor in Britain.

Yet the beauty of the bird is undeniable, the male in particular living up to its name *cyaneus*, with striking blue-grey plumage. That is echoed in the archaic name Ash-coloured Hawk, one of a great many alternative local and historical names listed by Watson, and seemingly applied also to the other grey harriers.

The females and immature males have both long been known as the 'ring-tails', because of the distinctive dark bands on the tail, which are of far less help in identification than the obvious white rump. They were once thought to be of a different species from the male. The ring-tail is built into the meaning of the name *pygargus* for Montagu's Harrier, which shares the same characteristic, as does the Pallid Harrier, *Circus macrourus*. In the case of the last species the more slender bird gives the appearance of a long tail, which is the meaning of *macrourus*. The ornithologist Colonel George Montagu (1751–1815) is commemorated in the one name, while 'Pallid' reminds us that the male is the palest of the three species.

In recent years a few immature Northern Harriers, *Circus hudsonius*, have been identified in Britain, including one long-stayer which I saw in Norfolk. I had earlier seen a female in Florida, when it had just been split from the Hen Harrier, but one of its older names in North America, the Marsh Hawk, tells that the bird' s habitat preferences may be closer to those of our Marsh Harrier. Indeed the pair which I saw in breeding territory in Nova Scotia were on marshy coastal meadows. The Florida bird was over mangroves, where it was wintering far from the northern breeding grounds which give it the vernacular name. The specific name *hudsonius*, after Hudson's Bay in Canada, reminds us that most breed between the Canadian border and the Arctic circle. The males tend to be darker than the Hen Harrier, and immature birds more rufous.

Marsh Harrier, *Circus aeruginosus*, is another species to show rusty tones, as its specific name suggests, but one of its older names, Bald Hawk, reminds us that the female is instantly recognisable with a cream-coloured crown. I have known that species for longer than any other harrier, since it became a regular breeder on the Essex marshes, to which I have returned throughout my adult life. My brother watches them from his house. However, it is worth a few cold fingers and toes to wait on a winter's evening at the Horsey Mere roost in Norfolk to watch Marsh Harriers, Hen Harriers, Merlins, Peregrines and Barn Owls while waiting for the Common Cranes. A count of over fifty Marsh Harriers there in 2012 convinced us that they have come a long way since their near-extinction in Britain in the 1960s. In international terms, this is now the Western Marsh Harrier.

I have never been so glad to see any harrier as I was when we were leaving Trinidad. We had looked for one all week, and had found about seventeen other raptor species, but no Long-winged Harrier. Then, within 200 metres of the airport entrance I yelled to our guide to stop. This species, *Circus buffoni*, is named for the French naturalist, Comte de Buffon (1707–1788). Unfortunately, *bouffon* is the French for 'clown', so I find it hard to put the two words together without a smile – but that smile gets even wider when I remember our luck at the airport.

LONG-WINGED HARRIER

Getting a buzz

As for our Common Buzzard, *Buteo buteo*, the name has onomatopoeic roots in the original Latin, and this comes into English circa 1300 as the Old French *busard*, gaining its modern spelling in 1616. Today it is the most numerous raptor species in Britain, having spread from the west towards the east at a remarkable rate over the past three decades.

Like so many of the older words for birds of prey, 'buzzard' had wider usage, being sometimes loosely used for 'kite'. It served in that general way for the bird we know today as the Honey-buzzard, but it was the discovery of a honeycomb in the nest of one which led Francis Willughby to create the name we now use. That name became current only after his early death, when his collaborator, John Ray, published their joint work. Its later adoption by Pennant ensured that it became the standard form. Though erroneous by today's understanding of the relationships of the species with other raptors (it is in fact closer to the kites), the name stuck. Today, the BOU prefers the spelling Honey-buzzard (with the hyphen), while our species becomes the European Honey Buzzard (with no hyphen) in the international English names. The generic name, *Pernis*, derives from the Greek *pernēs*, a type of hawk mentioned by Aristotle and Hesychius. The specific name *apivorus* means 'bee-eating', which suggests a very limited diet. However, BWPi reports the diet thus:

> *In summer and winter quarters, mainly nests, larvae, pupae, and adults of social Hymenoptera (wasps, hornets, bumble-bees); also, but more so in spring, other insects, amphibians, reptiles, small mammals, nestlings and eggs of birds, and occasionally spiders, worms, fruit, and berries.*

Buzzard or not, however, the similarity of the bird's silhouette in flight to that of true buzzards poses quite a challenge, even to experienced birdwatchers, particularly as it is still a rare species in Britain.

The scarce wintering species of true buzzard, the Rough-legged Buzzard, *Buteo lagopus*, gets its English name from its heavily feathered legs, while the Greek-derived *lagopus* echoes the generic name of the Willow Grouse and Ptarmigan in meaning 'hare-footed', i.e. heavily clad with a warm covering, as are the legs of a hare. I wonder if a hare notices the similarity as the buzzard's talons descend.

The Long-legged Buzzard occurs in southeast Europe. Its English name is a little deceptive: in reality the legs are very little different in length from those of Common Buzzard, so the name derives from the fact that more of the tarsus is exposed by the shorter 'trousers'. The scientific name, *Buteo rufinus*, refers to the golden-red of its tail.

Of hawks and 'hawks'

Birds of prey were fairly plentiful in Florida, not least because a good many prey species take refuge there in winter. I used to see both Sharp-shinned Hawk and Cooper's Hawk from time to time. They were quite difficult to tell apart, but there was usually some other species in the sky to give a size comparison. Both of these are true accipiters, which means that they are closely related to our Sparrowhawk and Goshawk.

The importance of that preamble is that I saw other hawks too, such as the ubiquitous Red-tailed Hawk, the Red-shouldered Hawk and the Short-tailed Hawk, but these were strictly hawks 'American-style'. Those three species are in fact buteos, of the same genus as the Common Buzzard, *Buteo buteo*. To understand this confusing detail, we need to see it is as a cultural issue. In fact, many Americans will still refer to the Northern Harrier as the Marsh Hawk, so the term 'hawk' tends to be a cover-all for raptors generally. And just remember that the term 'buzzard' is reserved popularly for the vultures, which are not true vultures, of course.

So we are again in the territory of being divided by the same language, though this is not such a big deal. Rather, you just get on and enjoy the birding. I have seen enough Red-tailed Hawks, *Buteo jamaicensis*, in both Florida and Canada to be convinced that they fill a niche very similar to our Common Buzzard, but the two areas produced two distinct subspecies, of which there are many spread widely through the Americas.

Red-shouldered Hawk, *Buteo lineatus* (meaning 'streaked'), is shorter-winged, and my encounter with one in the Everglades swamp forests showed me how easily it manoeuvred in dense tree-cover. I could also see clearly that the Florida race is much less streaked than the nominate one and that this immature bird showed no red shoulders. So much for descriptive names, in either language.

The small Short-tailed Hawk, *Buteo brachyurus fuliginosus*, which I found near Naples, Florida, was almost black underneath. *Brachyurus* means 'short-tailed' (compare *macrourus*, 'long-tailed', in Pallid Harrier), while the trinomial addition *fuliginosus* refers to the darker race of the species. This is a complicated matter, since most, but not all, of that race are dark, while a small proportion of the pale nominate race can be dark. Confused? I was later thrilled to see the pale race in Trinidad – and mightily relieved to know that location in both cases was an effective guarantee for each race.

Trinidad was rich in other so-called 'hawks': we saw Grey-lined, Common Black, Zone-tailed, Savanna, and White during a trip which found us some eighteen raptor species in a week.

Grasping at daws

Goshawk and Sparrowhawk are the British representatives of the genus *Accipiter*, and both names are used, with qualification, to name other related or similar species: in Australia, for example, I encountered Collared Sparrowhawk, Grey Goshawk and Brown Goshawk. In that context, the two British species become Eurasian Sparrowhawk and Northern Goshawk.

The Sparrowhawk is *Accipiter nisus*, Nisus being the son of Pandion II, and part of one of those legends where bad behaviour led to everyone being changed into birds (see the Osprey story). The word *Accipiter* derives from the Latin *accipere* (to grasp), which describes the fact that many of this group snatch their prey with their feet during the pursuit. Anyone who looks closely at the foot of a Sparrowhawk will see an extended central toe which acts as a devastating first hook for that job, as well as proportionately long legs to improve the reach. With that weapon, an element of surprise enables them to take prey as large as a Woodpigeon, which would average about twice the weight of a female Sparrowhawk. Jackdaws, Green Woodpeckers and the like are taken, though the much smaller male is limited to smaller prey.

The Sparrowhawk represents the extreme case of reverse sexual dimorphism (females larger than males) among British birds. (The size difference can be dangerous for the male: naturalist Keith Offord reported the case of one female which killed her hapless partner and fed him to their young.) Such dimorphism is more pronounced in bird-hunting species, though it is greatest in the Bat Falcon, *Falco rufigularis*, which we saw in Trinidad. Two theories underpin the reasons for this: the first suggests that a bird-hunting male is the prime hunter for the young, and that the female has evolved a preference for smaller partners which are no threat to her. The second is also illustrated by waders such as the Curlew, where different feeding capacities enable the males and females to use resources more efficiently by feeding across a wider spectrum.

Levant Sparrowhawk has a name which indicates its more easterly distribution. The Levant is a traditional term for the eastern Mediterranean region and originates from the Latin verb *levare* (to rise), the logic being that the Sun rises in the east. The scientific name, *Accipiter brevipes* (short-footed/toed), reflects the fact that this species lacks the extended central claw of the Eurasian Sparrowhawk, since it feeds on lizards and insects rather than on fleeing birds.

Of the two American accipiters, the Sharp-shinned Hawk, *Accipiter striatus* is worth further scrutiny. The odd vernacular name results from the fact that the front of the tarsus is sharply angled, rather than rounded as in most birds. The word 'shin' is strictly inaccurate: similar loose terminology is also found it the name 'thick-knee' when applied to the Stone-curlew. The streaking of the undersides (which is pronounced in juveniles) accounts for *striatus*.

I will return to the Goshawk a little later.

Of Peregrines and pilgrims

The word *Falco* describes a whole family of raptors with a curved and pointed wing-shape, which reminded observers of the scythe and sickle (in Latin, *falx, falcis*), hence 'falcon' via French. The most recent taxonomic reshuffle has removed falcons from their traditional place, which has long been next to the other 'raptors' (the eagles, hawks, buzzards and their allies). It seems odd to find the falcons now sitting on a new perch beween woodpeckers and parrots. In that place they are much closer to the owls, and not so very far from the pigeons.

It is no coincidence that colonies of Rock Doves will attract the Peregrine Falcon, *Falco peregrinus*: both species can now be found on the same sea-cliffs, and on the same cathedrals and office blocks, which are the man-made equivalent. Until the mid-twentieth century, Peregrines were heavily persecuted, on behalf of the military and with government blessing, in order to protect messenger pigeons used to convey vital communications. Today, illegal persecution of Peregrines places rogue pigeon-fanciers under suspicion.

Peregrinus has a medieval origin: Albertus Magnus, a twelfth-century Dominican scholar and bishop, and author of a treatise on falcons, *De falconibus*, studied a wild juvenile on its dispersal flight (peregrination) and attached the word *peregrinus*. Since a journey of pilgrimage was also known as a peregrination, and because the bird's black hood and dark mantle resembled a medieval pilgrim's travelling cloak, the name Pilgrim Hawk became attached as one of the popular names. *Peregrinus* started its life as an adjective, but we now often use its derivative, Peregrine, as a noun, dropping 'falcon' altogether. That is how language makes its own peregrination.

A number of other popular names existed, of which Duck Hawk tells why this large falcon hangs around lakes and marshes in winter. Other forms such as Game Hawk, Hunting Hawk, Spotted Hawk and Blue Hawk were also used, but the last name was also applied to the males of other raptor species.

As those examples show, the sharp distinction between hawks and falcons did not exist in popular usage. The word 'hawk' has its origins in the Germanic form, *hafoc*, of the Anglo-Saxons, and therefore had the wider usage among the majority of the population in the Middle Ages. The word 'falcon' belonged to the Norman French traditions of the ruling elite, for whom the sport of falconry was both important and exclusive. Flying a bird of prey was a hanging offence for those not of the nobility.

In fact, there was an entire technical language of falconry, of which the relevant information here is that, strictly speaking, the female Peregrine was the falcon and the male the tercel or tiercel. The phenomenon of reverse sexual dimorphism comes into play here, because the female is the larger and more powerful of the two, and therefore a different instrument from the smaller male. The origin of the name for the male lies in the Old French word *tierce* (third), which may relate to the one-third size difference, or, less probably, to the belief that the third chick was often a male.

Apart from the official persecution which the species suffered here during the twentieth century, the post-war use of DDT was almost enough to guarantee its extinction in Britain and elsewhere, since the poison had the greatest effect at the top of the food chain. After DDT was banned and Peregrines protected, the population picked up slowly until it has reached the point today when many consider it to be at a natural peak.

I now see them regularly, often several times in a year, and carry several special memories of them – the juvenile which three times missed a Lapwing over Pagham Harbour; the adult which lost its kill to a Marsh Harrier at Capel Fleet; and the bird which cruised high over the North Sea just after dawn, then flew low down the dazzle-strip of the rising sun to attack a wader roost.

The rivals

Two other sturdy falcons rival the Peregrine for strength: both are to be found in southeastern Europe and both have names full of curiosities.

The larger Saker Falcon, *Falco cherrug*, has a vernacular name which derives from the Arabic *saqr*, while *cherrug* is of Hindi origin, as mentioned earlier.

SAKER FALCON

Lanner Falcon takes its name from a Medieval French falconry term *lanier*, which first appeared in the thirteenth century. The origin of that is *l'anier*, meaning 'duck-hunter', which derives from an archaic word, *l'ane* (from the Latin *anas*, duck). The French suffix *–ier* implied 'trained to… '; it had parallels in other falconry terms such as *héronnier* (a falcon trained to hunt herons), and it is found reflected in later English terms such as 'retriever', 'setter', 'pointer' in the world of dogs and even 'mouser' among cats. The scientific name *Falco biarmicus* has a curious story. Temminck named the species in 1825, apparently in the mistaken belief that the word *biarmicus*, created by Linnaeus for the Bearded Tit/Reedling, referred to the two moustachial stripes (which this falcon has, in keeping with several others). The real meaning was in fact to do with Biarmia (Bjarmland), a region of Russia adjacent to the White Sea – which in turn reflected an error by Linnaeus (see Bearded Tit, later).

Of Hobby and hobbies

For me the Hobby, *Falco subbuteo*, shows the sickle-shaped wing better than any other falcon. The strange name, Hobby, derives from the Medieval French word *hober*, which meant 'to jump about'. This is a reflection of the bird's agility, and anyone who has watched one hunting dragonflies knows that it has breathtaking skills in the air. In falconry, it was the bird of the young page, since it was also capable of taking small birds. Its migratory movements follow those of swallows and martins, but it will take other species too, especially when feeding young. The one seen in Britain each summer is known more widely as the Eurasian Hobby, since there are three other closely related species.

HOBBY AEROBATICS

Subbuteo is composed of *sub* (near to) and *buteo* (buzzard). This seems an odd linkage, to say the least, but the concept seems to have been provoked by a word of similar meaning which was used by Aristotle. Its appearance in Merrett's 1667 list (spelt *subuteo*) suggests that it there meant the 'ring-tailed kite', which was undoubtedly the female Hen Harrier. That link is more akin to the strict meaning of the word. It appears that a line of misuse, which attached the word to Hobby, passed from Aldrovandus in 1599, via Ray 1678, and Albin 1731, to Linnaeus, who eventually formalised the name.

In 1947, Peter Adolph invented a new table-top football game, but the Patents Office considered 'Hobby' to be an inappropriate word to limit by law. Consequently Adolph resubmitted the second part of the bird's scientific name. For some decades Subbuteo was a top-seller of international fame.

For me, the hobby of watching birds has produced few more memorable days than one I spent at Grove Ferry in Kent one May morning in 2005. For some weeks the wind had been a stiff northeasterly, which had held up migration. The wind had suddenly dropped overnight, and we arrived in a confusing cool grey mist, as warmer air met the cold air. During the morning we became aware that Hobbies were everywhere, like flocks of large Swifts. The meteorological dam had collapsed and the delayed birds had streamed north in one mass. Given the mist, it was never possible to see just how many there were at one time, as birds appeared and melted into the ether, but one of my companions speculated that a three-figure count might not be unreasonable. It was certainly one of the largest influxes of recent years.

Red-footed Falcon, *Falco vespertinus*, is easily overlooked among Hobbies, with which it often flies. The English name holds no mysteries, though the odd immature bird can fail to have the necessary leg-coloration, as one of our Berkshire birds showed a few years ago. I recall that my first impression of the species was that the gaudy legs must have been made in a Chinese toy factory: they dominated my first view of a perched bird. The specific name *vespertinus* implies that this species is likely to be most active at dusk.

I have yet to see Eleanora's Falcon, *Falco eleanorae*, which was so named in 1830 by Alberto Marmora of warbler fame. Eleanora of Alborea was a fourteenth-century Sardinian warrior-princess, who decreed that the falcons were to be protected – presumably for the benefit of the noble pastime of falconry.

Of spells and spelling

The smallest of our falcons is the Merlin, *Falco columbarius*, but unfortunately there is neither legend nor magic in the name. Nonetheless, there is always magic in the sight of this little wizard streaking low in attack. The dull truth of the name is that it traces from an ancient Franconian German form, *smeril*, meaning a small hawk, and evolves via *esmeril* and *esmerillon*, to the Anglo-Norman *merillon*, and on to *merlin* – a series of different spells, one might say.

Strangely, given the long history of the species in Europe, the American form is the nominate race, and that may be at the root of the fact that it is called *Falco columbarius*, since it is also known in America as the Pigeon Hawk. Linnaeus himself first named this species in 1758, using a specimen sent from America, but initially called it *Aesolon columbarius*. In 1771, Tunstall described the European form and named it *Falco aesolon*. Subsequently the current name was adopted, but there is some suggestion that a split might yet occur between the Nearctic and Palaearctic forms: the argument is that there has been no gene flow between the Old World and the New for at least a million years.

Currently the subspecies breeding in Europe is *aesalon*, Greek for 'hawk', and a name linked historically with the Merlin. The Icelandic race *subaesalon* may also be found in Britain in winter. As to the specific name, *columbarius*, the implication is that the bird preys on doves, but that does not appear to be justified by the facts on this side of the Atlantic, where the Merlin normally preys on much smaller birds. In medieval falconry the Merlin was usually flown at skylarks, which it pursued high into the air: 'the merlioun, that payneth himself ful ofte the larke for to seke' (Chaucer). American sources do report that the Merlin takes Feral Pigeons, but the surprising thing about that fact is that the nominate race is also the smallest one. It is a curious fact that in falconry the Merlin was normally flown by noblewomen. Bizarrely, nuns of high-born stock would carry a Merlin on the wrist, and even into church – a bird of pray in that case.

My own favourite memory of the species was of one I found at Honeymoon Island in Florida in perfect February shirt-sleeve weather. My copy of Sibley confirmed that this was in fact a nominate Taiga bird, so to find it there was a bit of a shock. However, this is the one race which migrates across the Equator and, in any case, North America was then under a blanket of snow from the Carolinas northwards. Back at home, I have many vivid memories of watching Merlins streak through winter marshes in pursuit of hapless Dunlin and the like.

A Kestrel for a knave

The title of the 1968 novel by Barry Hines recalled one of the rules of falconry etiquette, by which a person of relatively low status might fly nothing more than a Kestrel. The truth is that *Falco tinnunculus* pursues inconsequential prey (small mammals and insects), which would bring no kudos to the person flying it.

KESTREL

The word Kestrel has its roots in the French *crécerelle,* from *crécelle* (a rattle), which refers to the rattling call uttered in courtship. In like vein, *tinnunculus,* the Latin for Kestrel, derives from *tinnulus* (shrill sounding).

Kestrels are a subset of the falcon family, of which there are several species, and I have seen three others on a number of occasions. The sharp-eyed among you may have noticed that Lesser Kestrel, *Falco naumannii,* is the only species in the *Collins Bird* Guide to be depicted with a specific and recognisable building, which is the Giralda of Seville Cathedral. When the Spanish built the cathedral on the site of a Moorish mosque, their facade provided a warren of Gothic ornamentation in which Lesser Kestrels nest. They also recycled the spectacular Moorish tower, from which I observed no fewer than eighteen Lesser Kestrels circling – just as in the book. And, by the way, while Johann Naumann's name is given to the bird, the reputed architect of the Giralda, Al Geber, gave *his* name to algebra.

My first encounter with Australian Kestrel, *Falco cenchroides,* was a bird hovering over a coastal golf course at Newcastle, NSW, looking very like a Common Kestrel (but paler). Its alternative name is Nankeen Kestrel. The word also occurs in the name of the Nankeen Night-Heron and derives from the pale rufous colour of a traditional Chinese cotton trade-fabric of that name. The scientific name *cenchroides* simply states that the bird 'resembles a kestrel' (which is *kenkhris* in Greek).

The American Kestrel, *Falco sparverius,* gets its specific name from a medieval word for Sparrowhawk, which was transferred to the falcon as a reflection of its small size. I recall that the year after finding the Taiga Merlin, we returned to Honeymoon Island and found an American Kestrel sitting on exactly the same snag. Curiously, it was unlike the pale local race, *paulus* (small), which I saw regularly in Florida, but was the dark-streaked nominate form which I had seen in Canada. Again winter weather prevailed further north.

Of Gyr and Gos

It seems fitting to deal with Gyrfalcon, *Falco rusticolus,* and Goshawk, *Accipiter gentilis,* in tandem. For one thing they were both considered to be birds of great stature in falconry: indeed the specific name *gentilis* implies that the hawk is a bird for the nobility. Not least, Aldrovandus reminded us that the hawk's truculence commended it as the chosen symbol of Attila the Hun, no less. But I feel that there may well be a further link between the two birds...

Gyrfalcon was, in medieval falconry, the bird of the kings. It is larger than Peregrine, though flies less high in pursuit of prey. The name first appeared in English as *gerfauk* in the early fourteenth century. That word has a complicated pedigree, since it came into English via a French form which had derived from Old Norse *geirfalki.* In that form, *geir* (spear) denoted excellence, but the *falki* element was taken from an Old German adaptation of the Latin *falco.*

All of those complications were compounded by Pennant's adoption of the current spelling, in which he noted Aldrovandus's 1603 use of *Gyrofalco.* This form had rationalised the Norse element of the word into the concept of *gyro* (circling), which was tagged onto the bird's apparent habit of circling to look for prey. That is more than a touch of wishful thinking, or even of arrogance, since the classicists of the Renaissance onwards often saw northern cultures and Germanic languages as barbaric and untutored. Pennant himself exhibited this trait fairly frequently in his preference for words which he deemed 'superior' to the Anglo-Saxon forms. Since the time of Pennant, both Gerfalcon and Gyr Falcon have been used, the latter still being the preferred form of the BOU, but Pennant's version, Gyrfalcon, is the accepted international form.

The specific name *rusticolus* means 'country dweller', but that seems a feeble choice for a bird of the remote High Arctic and rugged cliffs. Recent studies by Burnham & Newton have shown that the bird in fact spends most of its winter at sea on pack ice and icebergs, in pursuit of auks.

The Goshawk derives its common name from the term 'goose hawk', which is first found in Old English as *goshavoc.* Lockwood doubts that they were flown against geese. This view seems reasonable, since any birdwatcher will know that Goshawks have evolved for ambush hunting. BWPi contains the following note: 'Open country exploited only when it lies near more or less extensive woodland, and when hedgerows or wind-breaks provide good cover. Favours areas where woodland interspersed with fields or even wetlands.' Only the last word suggests that it might expect to encounter geese, and indeed there are no mentions of goose species among the prey items quoted in studies by BWPi, so there must be another explanation.

Lockwood in fact suggests that the Peregrine might fit the name goose-hawk better, and links that idea to the Old High German *ganshabuh,* denoting 'a species of falcon, almost certainly the Peregrine'. There is in fact some evidence that Peregrines do occasionally take geese. However, it seems reasonable to speculate that the Gyrfalcon, which is known to take geese more frequently, could well have have been the original *ganshabuh* and that somewhere along the line the names of the largest falcon and the largest accipiter were confused.

BY LAND AND WATER: CRAKES, RAILS AND GALLINULES

The term 'waterfowl' used to make a good general term for this group of birds, but I note that the IOC list has now used that term to replace 'wildfowl'.

Between reeds and weeds

The words 'rail' and 'crake' have been used in parallel for many centuries, as the stories of Corncrake and Water Rail show. Today they are applied to two groups within the family Rallidae, the rails being generally characterised by longer bills, and the crakes by shorter, stubbier bills. The rails and the crakes are so broadly similar that Lockwood reports that an old superstition had the Water Rail as the winter persona of the land-based Corncrake. That is not as silly as it sounds: Corncrake was once more widespread throughout Britain as a summer visitor and was certainly well known for its rasping voice, while Water Rail tends to be more visible at the edges of reed-beds in winter, long after Corncrakes have moved back south.

WATER RAIL

The name Water Rail, *Rallus aquaticus*, has remained unchallenged in form since used in Latin by Aldrovandus in 1603 and, in the English form, by Mouffet in 1655. Ray followed suit in 1678 and Pennant in 1768.

The name Corncrake had a bumpier road towards its current acceptance as the formal vernacular name for *Crex crex*. It existed from at least 1455, but in 1678 Ray created the term Land Rail from the Medieval Latin *Rallus terrestris*, and that form was favoured by many until the end of the nineteenth century. In 1776, Pennant created the term Crake Gallinule in an attempt to give the bird a suitably scientific vernacular name: it proved to be inaccurate anyway, and soon fell out of favour. Bewick reinstated the word Corncrake in 1797, and this gained ascendancy, to be favoured by the BOU in 1883 and then Hartert *et al.* in 1912. It was used by Coward in 1926 in the form Corn-Crake. The current IOC list goes a step further with Corn Crake, possibly because some ornithologists prefer vernacular names to follow the two-word pattern of the scientific forms.

As scientific names go, *Crex crex* is beautiful in its simplicity and its effectiveness. The simple repetition of an echoic word coined by the Greeks as *krex* tells us exactly what to listen out for. For once, the scientific name has a homely feel to it, and is as good as Cuckoo. In my own case, though, it was my failure to hear one of these, when it was said to be keeping some cottagers awake at night, that sent me straight to the hearing clinic next day, bitterly disappointed to have missed the sound. The word 'crake' itself refers to the rasping call, and its variant reappears in the word 'drake'. The word was also used in northern dialect for the crow, a derivative of Old Norse *krāka*.

The origin of the word 'rail' is also in onomatopoeia: it starts life in the Latin *rascula* (a rasping noise), and travels into English via Medieval French. In that way it is related to the Modern French *râle* (a death rattle). Rails and crakes are slim, a fact which enables them to slip between grasses and reeds, but the use of the same word in 'thin as a rail' is merely coincidental, for that expression starts life in another Latin word, *regula* (rod).

We saw Mangrove Rail, *Rallus longiristris* (long-billed), in Trinidad, where the subspecies is peladromus, meaning 'mud-running. In all, that is a pretty good combination of names which together tell us a useful set of things about the bird we saw slipping through mangrove roots at the edge of Caroni Swamp.

The Spotted Crake's scientific name, *Porzana porzana*, derives from a Venetian name for a small crake, which was used by Buffon and adopted by Linnaeus. The bird is so secretive that there were no precedents for its name when Pennant described it using the word 'spotted', but after being known as Spotted Gallinule and then as Water Crake, it was finally named by Shaw in 1824 in the form which we now know. My own experience is of a bird which can be so frustratingly difficult to see, and then only ever on passage. I was once at Dawlish, in Devon, when I was told of one at Topsham. Arriving in the hide, I found another birder there. His tip was: 'Watch that hole in the reeds and wait about five minutes'. It was true: the bird popped out every twenty minutes or so like a cuckoo in a clock.

Another of my favourite encounters was with the even-smaller Baillon's Crake, *Porzana pusilla*, this time in an Australian urban park. The word *pusilla* (tiny) underlines the fact that it makes a small target, even when not hidden by reeds. On that occasion I was standing at the edge of a thin strip of reeds, craning for a view of an Australian Reed Warbler which was singing its heart out. Mary alerted me to the fact that the Baillon's Crake was walking almost over the toe of my shoe, so close that by the time I had the camera focused I missed the best shot. Here it was the subspecies *palustris*, a word which refers to its marsh habitat. And, by the way, Louis-Antoine Baillon was a nineteenth-century French collector. With a small western European population, this species is also a vagrant to Britain, as is the slightly larger Little Crake, *Porzana parva* (small).

Buff-banded Rail, *Gallirallus philippensis*, showed up a number of times in Queensland. *Gallirallus* means literally 'hen-like rail'. This species has a wider distribution through the Pacific region, as the name *philippensis* implies: there are in fact well over twenty subspecies, and the Australian one is *mellori*, named after the Australian ornithologist John White Mellor, a founding member of the RAOU.

When 'moor' is less than satisfactory

A third grouping within the Rallidae is represented in Britain by the Common Moorhen, *Gallinula chloropus*. Today the word 'moor' tends to imply an area of open upland, but originally its sense included marshland too, as evidenced in the name Moor Buzzard, which was once applied to the Marsh Harrier. The name Moorhen had been in use since the thirteenth century, alongside Waterhen (which name survives in Ireland), but in his usual style Pennant rejected a word of common usage in favour of the Latinate term Common Gallinule. That is based on *gallinula* (a little hen), from which derives the generic name of the moorhens. 'Gallinule' has survived in the Americas, since it seems that there is little sympathy with the root form, particularly given the change in the value of the word 'moor' since its origins. Yarrell reinstated Moorhen in 1843 and that has stuck on this side of the Atlantic, and for a time was the agreed international form. However, in 2011 the North and South American committees voted to split the American form as a separate species, *Gallinula galeata*, and the species now appears as such on the IOC list. The nominate form, *galeata* (helmeted) is the South American race, which we saw in Trinidad, while the North American race, which we saw in Florida, is *cacchinans* (laughing). These names seem to have been chosen with particular care, since they each refer to key features which are evidenced to justify the split: the shape of the head-shield and the voice, which are both significantly different from *chloropus* (which, by the way, means 'green-legged').

There is no problem with the Australian Dusky Moorhen, *Gallinula tenebrosa*, since it is a distinctly different beast, as large as the Australian Coot (of which more in a while). The word *tenebrosa* (dark, gloomy) merely echoes the Dusky. We also saw two flocks of its relative, the Tasmanian Nativehen, *Tribonyx mortierii*, which is larger still and distinctly greener. The sheer size of the bird justifies its comparison to a domestic fowl, and it is endemic to Tasmania, so native in that sense. The specific name was given in the nineteenth century by du Bus de Gisignies to honour a fellow Belgian scientist, Barthélemy du Mortier (1797–1878). *Tribonyx* means 'worn claw', which reminds me that Wikipedia cites a modern vernacular form which caught my eye: the bird is flightless and runs very fast from danger, since it spends large amounts of time foraging in open grassland. It apparently has the soubriquet 'turbo-chook'. Only an Australian could have come up with that idea. For the uninitiated, 'chook' is a chicken. My father once noted a sign outside a fast-food shop in New South Wales stating: *Fried Chook and Take-away Tucker*. The language is known as Strine.

Purple pets or purple pest?

We came across the Moorhen-sized Purple Gallinule, *Porphyrio martinicus*, in Trinidad, having failed to find one in Florida. In this case the specific name is derived from Martinique in the Lesser Antilles. There has been frequent confusion of the vernacular name with the much larger Purple Swamphen, *Porphyrio porphyrio*, partly because the American species is an occasional vagrant to Europe and West Africa, and because the Old World species has become established in Southern Florida as an escape. *Porphyrio* is derived from the Greek name of the species, *porphurion*, from the Greek for purple.

As a very good example of the 'drifting mines' which I mentioned earlier, the dozen or more former races of Purple Swamphen have very recently morphed into six species. The all-purple nominate race, which we saw in Andalucía, has now become the Western Swamphen, so out should go all the confusion with the gallinule. The former Australian race *melanotus*, meaning 'black' (it is black-backed), is now a full species as the Australasian Swamphen. Most, though not all of the Florida escapes are of a Caspian Sea form, *poliocephalus* (grey-headed), and have the added distinction of being the equivalent of the Ruddy Duck in Britain – subject to eradication orders.

Swamphens were well known in the Ancient World, and were prized as pets of some prestige. Curiously, something similar occurred in Maori and Polynesian society, and that fact relates to my own first encounter with the Australasian species in a park in Canberra. One particular individual was so used to being fed that it came up to beg like a dog. I described it in my diary as the most unsatisfactory 'in your face' first encounter with an exotic bird. Fortunately I have subsequently seen a good many of them in truly wild places.

AUSTRALASIAN SWAMPHEN

Coots and cuter Coots...

In the world of coots the picture is the reverse of that for the Moorhens: the American species, *Fulica americana*, is indisputably distinct from the Eurasian one, but the Australian bird is currently considered a race of the Eurasian species.

The word 'coot' is onomatopoeic, based on the sharp call of the bird. In international terms the British species is the Eurasian Coot. Its scientific name, *Fulica atra*, means 'dull black Coot' in Latin. The Red-knobbed Coot, which occurs in Spain – but avoids me! – is *Fulica cristata* (crested), a reference to the two red blobs which occur at the top of its frontal plate.

The race of Coot found in Australia is *Fulica atra australis*, which is considerably smaller than the nominate race seen in Europe. Bizarrely, to English eyes, it is actually smaller than the Dusky Moorhen. Though it is referred to in Australian handbooks as Eurasian Coot, the name Australian Coot is clearly more suitable and is certainly used in New Zealand to refer to the population which has established there in recent years by natural migration to fill the space left by two extinct species. Although a number of other Australian forms (White-headed Stilt, Eastern Osprey, Eastern Cattle Egret and, lately, Australasian Swamphen) have recently been split, this one is not yet at the starting gate, though it seems a good candidate.

OF DRAINS, PLANES, STONES AND CRANES

For British birdwatchers the formal English name of the Common Crane, *Grus grus*, appears to start with an outright lie. Cranes are certainly not common in Britain, although the tiny East Anglian population has kept the species on my lists in recent times, and the reintroduction project on the Somerset Levels has been designed to hasten the comeback of a bird that was once widespread, but largely lost to the drainage of wetlands during the eighteenth-century Agricultural Revolution. They are, of course, much more common in mainland Europe, and have been recorded widely since the time of Aristotle.

COMMON CRANE

As Lockwood describes precisely, the word 'crane' has roots almost as ancient as language, in that it starts its life in an Indo-European form which has fed many languages and has evolved through the Germanic lines towards the version we now use. The origin of the Latin *grus* can be seen in the Greek *géranos*, which is much closer in sound to the word 'crane'. Sound is in fact important here, since the root names are almost certainly onomatopoeic. The word *géranos* itself is echoed in the obscure Greek legend of a Pygmy Princess, Gerana, who offends the goddess Hera and is turned into a Crane.

Both Aristotle and Pliny discussed the species, and both left elements of pseudo-folklore which were perpetuated by scholars who recycled them without making observations in the field. Both spoke of stones, Aristotle of regurgitated stones of interest to the alchemist, and Pliny of a sentry bird holding a stone which it would drop if it fell asleep. Behind the nonsense is some truth: the birds do take stones into their gizzards to aid with digestion, and, like geese, sentry birds do keep watch.

Other crane species are part of culture and legend around the world, their display dances in particular admired and imitated in dance and art (see my later note about the Brolga Crane). Sadly, very many of the cranes are in some decline or under real threat, and it was interesting to see the Russian President involved in flying in a 'parent-bird' microlight during a crane reintroduction project in Asia a few years ago. The technique has been used successfully in America to teach hand-reared Whooping Cranes, *Grus americana*, to migrate. Sadly my Florida visits never managed to produce either that or the Sandhill Crane, *Grus canadensis* – but I also have a few other species to go back for. We were luckier in Australia, where we saw Brolga Cranes, *Grus rubicunda*, in both New South Wales and the Northern Territory.

The extreme southeast corner of Europe is in the breeding range of the beautiful Demoiselle Crane, *Grus virgo*, whose names appropriately refer to a 'maiden' in French and Latin.

NEVER TOO LATE BUT OFTEN GAME: BUSTARDS

The word 'bustard' arrived in English as a combination of two French forms, *outarde* and *bistarde*, which Lockwood traces through two evolutions, the former from a Gallic development from Latin and the latter from an Italian form which was borrowed into French. The Latin root itself is curious, since it was originally *avis tarda*, which appears to mean 'slow bird'. Since the Great Bustard, *Otis tarda*, would be in line for the European title of *turbo-chook*, there has to be an explanation for this apparent nonsense, and it was provided long ago by Pliny. He explained that a local Iberian word for the bird had been corrupted into the nearest available Latin word, regardless of the meaning. The scientific name *Otis tarda* is curious in itself, since the Greek *Otis* referred to a 'long-eared species' of bustard (compare *Otus* in owls), though the prominent 'whiskers' of this species have to suffice. The word *tarda* more correctly recalls the Iberian origins rather than the misleading Roman Latin. My only encounter with one was a long-staying bird which occurred in Wiltshire a few years ago. The exciting thing about this bird was that it had neither tags nor rings, so was apparently not part of the reintroduction project.

Macqueen's Bustard, *Chlamydotis macqueenii*, has been recorded only very rarely in Britain. Its generic name derives from the Greek for a horseman's cloak (clearly for its contrasting mantle) plus the above word *otis*. General T. R. Macqueen presented the first specimen to the British Museum in about 1832.

Australian Bustard, *Ardeotis australis*, was a highlight of our trip to Kakadu, even if fairly distant. The generic name marries *Ardea* (heron) and *Otis* (bustard) together in a way which reminds us that this is a rangy bird.

I have never yet seen a Little Bustard, *Tetrax tetrax*, but the scientific name is a Greek word for an edible game bird mentioned frequently by assorted Greek authors. It is significantly similar to *tetrix*, which appears in the name of the Black Grouse.

ALIAS SHOREBIRDS AND PEEPS: WADERS

As in the case of wildfowl, birds of prey and game birds, our collective term 'waders' is less than perfect, since some of them do not do a lot of wading. However, the American term, Shorebirds, has the same sort of problem, because some of the species covered in the blanket term are rarely, if ever, found on shores. However, both terms exist, and it is likely that a publisher will have to choose one term or the other according to the target market. I have no great passion about this issue, but it has to be remembered that the Americans also use the term 'waders' for what the British would tend to call 'herons, egrets and their allies'. There is no doubt that it matters little in reality, provided we are alert to those differences of usage. I tend to think that the American term 'peeps', for small sandpipers and plovers, is an attractive idea, since it is more concise than any alternative which you might hear uttered in an average British hide.

So this next section is about the names of what the British would refer to as waders.

Oyster pie and mussels: oystercatchers

EURASIAN OYSTERCATCHER

Oystercatcher is an odd name in a number of respects. Firstly, it is rare, if not unique, in being an American import, coined by Catesby in 1731, after he had watched the American species, *Haematopus palliatus,* feeding on oysters. Prior to that, the Eurasian species, *H. ostralegus,* was often known as the Sea Pie, the shore equivalent of the Pie (Magpie). Lockwood notes the use by Hartlepool fishermen of Mussel Cracker, which would have made a more apposite name, since the Eurasian species favours mussels. There has been speculation that oysters were once more plentiful, but it seems more likely that the two separate species on two continents had evolved as specialist feeders on different molluscs. The irony is that it is the Eurasian species which bears the specific name *ostralegus,* meaning 'oyster gatherer', while the American species, *palliatus*, has to make do with a name meaning 'cloaked'. Seen well, the American Oystercatcher does have a paler greyish-green mantle at all ages, which is akin to that of the juvenile Eurasian bird. I have to confess that I had to look hard to spot that feature when I first saw them in Florida, but was sure that the call was just not the same. The word *ostralegus* actually makes more sense than the name Oystercatcher, since the latter implies an active and evasive prey. It has long been known that some birds open mussels with a blow (hence Mussel Cracker), while others prise them open. For this, the bills are adapted in shape. Naish quotes studies which also show that bill adaptation allows some to probe for worms, and that adaptation is reversible over a period of time to allow for resource polymorphism, i.e. a choice of diet. It is not always evident that the bills of some species change in this way: the Starling's bill is black in winter because it is strengthened with melanin for a greater amount of probing in the soil. The seasonal change of the Puffin's bill is a better-known example.

There are several other oystercatcher species around the world, all linked by the generic name *Haematopus*, meaning 'blood-foot' and referring to the bright pink legs. I was lucky enough to see two together on the beaches of Tasmania, Sooty and Pied, the latter looking very similar to the American and Eurasian pied forms.

Much more than a logo: Avocet

PIED AVOCET

Avocet is yet another of Pennant's preferences. It comes from a French form, *avocette*, which in turn derives from a Venetian word, *avosetta*, of unknown origin. However, while some speculate a link to the garb of the Italian *avvocato* (lawyer), the *Dictionnaire Etymologique* suggests a possible Latin root in *avis sitta*, 'a sort of magpie'. The word was introduced by Aldrovandus in 1603, and used by both Charleton and Ray in the late seventeenth century. Linnaeus recorded the name *Recurvirostra avosetta* in 1758. Again Pennant chose it as a more genteel word than the available popular forms, initially as Scooping Avoset and later as the single word Avoset. In this case I am inclined to sympathise with him, since most of the names in use were less than attractive. Could any of us imagine that the RSPB's logo bird could be better known as a Crooked Bill, a Cobbler's Awl, a Scooper, a Barker, a Clinker, or a Yelper? As for this penchant of Pennant for things French, it must be remembered that the manners and foibles of the French court had dominated polite British society after the Restoration of Charles II in 1660 and still permeated it right up until Pennant's old age. The arrival of the spelling with a *c* is apparently down to an erroneous transcription by Brisson, who created the French form *avocette* in 1760. Because there are other avocet species, the European bird is now Pied Avocet in international English. *Recurvirostra* simply means 'upturned bill'.

Teetering on the brink: stilts

The extra long legs of the Black-winged Stilt, *Himantopus himantopus,* have resulted in it having a common name based on the wooden poles used to elevate workers doing high-reach jobs, and in circus acts. The scientists, however, reverted to ancient Greek, using a word which described legs which folded like straps (*himas, himatos,* strap, and *pous,* foot). Curiously, the unrelated and beautiful Australian Pratincole goes by the name *Stiltia isabella,* the generic name simply being Latinised English. As for *isabella,* see my earlier comment. It is the longest-legged of the pratincoles and was a delight to see in the billabongs of Kakadu.

The traditional problem for the stilts has been uncertainty about which forms were species and which subspecies. A few years ago the Americans led a move to separate the Black-necked Stilt as *Himantopus mexicanus,* and very recently the locally named Pied Stilt of Australia has been formally split as White-headed Stilt, *H. leucocephalus.* The scientific name says the same thing as the English. I found some of these birds in the Hawkesbury Wetlands and saw more in Kakadu: they seemed to show a short black 'mane' of neck feathers. The American species first appeared for us in Myakka, Florida, was numerous at Merritt Island, and was present in Trinidad at the much-less-glamorous Trincity Sewage Works.

Of falsehood, winnowing and helmets: lapwings

One good reason for Lapwing to be included in my title is that it appears to have had more vernacular names than any other British bird. That is surprising, given that the species is on the red list of endangered species in Britain. But the truth is that Lapwings were both more common and more widespread in times past, so it was inevitable that such species tended to gain the greatest number of popular names. The converse of that is the secretive Spotted Crake, which apparently had no common name until the scientists got hold of it: a little-seen, insignificant, and rare bird like that simply didn't get noticed. But the Lapwing was obviously a different case: it was numerous, striking in plumage, flamboyant in flight and noisy to boot, so it was well known. It was also a benefit in the fields, since it would feed on the creatures of the soil rather than on the crop, thereby performing a service in pest control. But it was also a welcome resource: I recall my father telling me that, in the 1920s, he went out with his brothers during the nesting season to collect the eggs of Lapwings, Coots and Black-headed Gulls. They helped to feed the large family. This practice was centuries-old and did not undermine populations with healthy numbers. It was the subsequent advent of intensive agriculture, the disappearance of much grassland, and the use of chemicals which did the damage: the breeding population of Lapwings in Britain has declined so dramatically in a few generations that we are easily surprised by the numerous names it was given in the past.

NORTHERN LAPWING

The word Lapwing itself is firmly rooted in Old English *lǣpiwince,* a word documented as early as the eighth century. Curiously it originated in the linking of two older Germanic words, both loosely signifying 'a crested bird'. Variants such as Lapwinch, Lapwink and Lappinch existed in various dialects. Other popular forms such as Wick, and Wipe, and variants of these, reflect the crest. Peewit and many variants were based on the call. The list is impressive, but far too long to include here. Pennant, once again, was the man who made the choice which stuck after 1768.

It was the crest that got the Lapwing into some trouble in the Middle Ages. In the fourteenth-century *Parliament of Fowls,* Chaucer described the bird as 'the false lapwynge full of treacherye'. This, to a modern birder, will not seem to correspond with the facts, and indeed one of my most treasured memories is of the valiant and pitiful attempts of a bereaved Lapwing to drive a Peregrine off the corpse of its partner. What we are dealing with is a different sort of language – the language of allegory. Endowed characteristics became symbols to decipher in the allegorical mind-games played in the Middle Ages: such symbolism still lingers today in the trade slogan: 'say it with flowers', and the forget-me-not says it clearly. The explanation of Chaucer's strange accusation is linguistic and literary, rather than ornithological, and it is linked with Greek legend in a curious way. In the original legend, King Tereus lusts after his sister-in-law, is outwitted, and then pursues Philomela and Procne before they are all turned into birds by the gods. In Ovid's Latin version, Tereus becomes a Hoopoe, and somewhere along the line the Lapwing is the victim of mistaken identity when the Latin noun *Upupa* is used adjectivally to describe another crested species, the Lapwing. This is well illustrated in the Modern French, where the former is *la huppe* and the latter *le vanneau huppé*. Of course, the Hoopoe was nothing like as well known in England as the Lapwing, and this was literally a question of the cap fitting in the form of that crest. Hence 'the false lapwynge full of treacherye'.

It also seems likely that the concept of treachery enters the collective term, a 'deceit of Lapwings', and what is apparently a corrupted version of this, a 'desert' of Lapwings. This reputation may also have underpinned a much later dislike of the bird in Scotland, after clansmen hiding in the heather were betrayed to the Redcoats by anxious nesting Lapwings.

However, all that is past and the bird is now the Northern Lapwing, *Vanellus vanellus*. But where did that word *vanellus* originate? Apparently it is the diminutive of the Latin *vannus* (a winnowing fan). You are none the wiser? The truth is that we don't do a lot of winnowing these days. The action takes places on the threshing floor, of course, and the winnow is used after the flail. But life moves on, and the vocabulary of the past becomes obscure to a point that the last sentence may be near-meaningless for many people living now. In modern agriculture, all such action takes place within the combine harvester, when the grain is separated from the husk (threshed), and the resultant chaff blown away (winnowed). To return to *vanellus*, the broad wing of the Lapwing resembled the large hand fan used in winnowing. For British birders at least it is a matter of separating the wheat(ears) from the chaff(inches).

Today, our Lapwing is the Northern Lapwing since lapwing has now taken on a wider value as an international generic name. Sociable Lapwing, *Vanellus gregarius*, and White-tailed Lapwing, *V. leucurus*, are both rare vagrants to Britain and both have scientific names which echo the vernacular names.

I encountered Southern Lapwing, *Vanellus chilensis*, in Trinidad. The specific name *chilensis* (from Chile) reminds us that it is a widely distributed South American species. In the case of the Trinidad birds, these were *V. c. cayennensis*, a more northerly subspecies, whose trinomen comes from Cayenne (French Guiana). In like vein, the Australian species, Masked Lapwing, is *Vanellus miles*, but this one comes in two very different-looking subspecies. The southern race is nominate and wears distinctive wattles and a black hood (which justifies the specific name meaning 'soldier'), but the northern race, *V. m. novaehollandiae*, has a much more pronounced 'helmet'.

At which point we are quite a long way from the eighth-century usage of *lǣpiwince*.

Plain wrong, doubtful or dotty... plovers

Both 'plover' (via the Old French *plovier*) and the Latin *pluvialis* are derived from the Latin word for 'rain', as it was thought that some plover species called before rain.

The Grey Plover, *Pluvialis squatarola*, gets its scientific name from a curious Venetian name for it, *sgatarola*. For once a bit of creative imagination went into the naming of the Golden Plover, *P. apricaria*, since it apparently sunbathed to get that gorgeous golden colour: the Latin *apricari* means 'to bask in the sun'. In international terms we need to add Eurasian to distinguish it from the American and Pacific species.

The smaller plovers belong to the genus *Charadrius*, which is Late Latin for 'a yellowish bird'. The origin of that is the Greek *kharadrios*, a drab bird living in ravines (*kharadra*) and thought to be Stone-curlew. None of that relates at all to the normal habitat and behaviour of these birds, so it seems to be a spare name which happened to be in the drawer at the time. Astonishingly, this nonsense is perpetuated in the specific name of the most familiar of these, the Ringed Plover, *C. hiaticula*, since its specific name literally means 'cleft-dweller'.

From wholly wrong to somewhat dubious, but in a wholly different way... In using the Latin for 'doubtful' in the name of the Little Ringed Plover, *Charadrius dubius*, Scopoli implied (rightly, for many birders) that it was easily confused with its larger cousin. The latter should technically be called the Common Ringed Plover – presumably just in case we are tempted to write about it as 'a cute little ringed plover'. Such an example underlines the case for using capitals for bird names. Scientists do not approve of the practice, but it is in fact helpful, I feel.

The Eurasian Dotterel, *Charadrius morinellus*, seemingly owes both its scientific and its vernacular name to its eccentric (dotty) behaviour. *Morinellus* relates to the *morus*/booby elements of the Gannet names: too trusting of humans and therefore 'stupid'. Lockwood, however, notes that a Norfolk name, Dot Plover, suggests that the name is based on a single note at the end of a soft, fluting call, later embellished with the suffix *–erel*. That was often the case with names based on the call – compare Cockerel and the archaic Whimmerel (for Whimbrel). He feels that it is a coincidence that the concept at some point collided with the meanings of 'dotage' or 'dotty' to give rise to the scientific form.

EURASAN WHIMBREL

We now need the word Eurasian in front of Dotterel because other plover species bear the name. One of my personal favourites is the Black-fronted Dotterel of Australia, which I have seen several times. It breaks ranks by having its own genus, *Elseyornis*, based on the name of J. R. Elsey, a nineteenth-century Australian naturalist-explorer: the name means 'Elsey's bird' and was coined by Vieillot. Its specific name, *melanops*, supports the English name: it means 'black-faced'.

Other *Charadrius* species appear in Britain as rare vagrants from much further east. Caspian Plover's name recalls the Caspian Sea, but its scientific name *C. asiaticus* reflects a wider range, while Lesser Sand Plover is *C. mongolus* (Mongolian). Greater Sand Plover's name *C. leschenaultii* reflects this concept only indirectly, since it was named after a Frenchman, Leschenault de la Tour, who collected in the East Indies and in India and Ceylon.

The one vagrant from the Americas is the Semipalmated Plover, *Charadrius semipalmatus*, whose names refer to the partial webbing of its feet. I found in Canada that this species is so similar to our Ringed Plover that it was easy to overlook its smaller size. It was a useful yardstick, however when I chanced upon a mixed flock in Florida which contained Snowy Plover, *C. nivosus*, and the much rarer Piping Plover, *C. melodus*. Neither of those carries any mystery in the scientific names, each of which conveys the same sense as the English. Snowy Plover used to be considered of the same species as Kentish Plover, but was split a few years ago. I searched the beaches of Florida most diligently for Wilson's Plover, *C. wilsonia*, but had to wait until my trip to Trinidad to see one. This species is one of many named after Alexander Wilson.

(Kentish Plover and Killdeer are dealt with elsewhere.)

Tide-in knots: small sandpipers

For the origin of the generic name of the *Calidris* sandpipers we have to go back to Aristotle, to find that he used the word *kalidris* or *skalidris* to describe a grey waterside bird.

Dunlin might easily be called the 'yardstick' sandpiper, since it is the familiar common wader against which all other small waders are judged: Knot are therefore 'bigger than' and Little Stint 'smaller than' the Dunlin. Not that its real name means anything important: 'dun' is the colour and the diminutive *–ling* is employed in a truncated form to make the name: in short, it is literally the wader form of a 'little brown job' (compare Dunnock, later). Its scientific name, *Calidris alpina*, tells us that it is a mountain breeder: 'high country' or 'high latitude' might be the better interpretation.

DUNLIN

It follows logically that if one bird was called by a name meaning 'little dun', there had to be a bigger 'dun', which was in fact the name of the Knot in some parts of the country. 'Knot' itself has a pretty well-known story. The eleventh-century English King Canute (Knut), some legends tell us, demonstrated that he had no power over the tides. One version is that the Knot got its name because it too frequents the tidal zone, but I feel that the Sanderling would have been a better candidate, since it scuttles so much more dramatically along the water's edge. On the other hand, those huddled flocks of Knot do look a bit powerless as the tide rises, forcing them to vacate the beach to roost somewhere above the tideline. But there is another, equally credible version, in that there are those who maintain that the Knot was a favourite of King Canute. Not that he spent a great deal of time with his bins and notebook: he was said to find Knot very tasty, fattened up and served with bread and milk. It is not such a tall story when we find that Muffet, in 1595, states that 'Godwits are sold at four nobles to the dozen' (Bircham). Even today,

Snipe and Woodcock are considered by some to be edible 'game birds'.

Today, because the Great Knot, *Calidris tenuirostris* (slender-billed), has been added to the catalogue of waders, we should refer to ours as the Red Knot. Not that we see it in its breeding plumage in Europe that often. When I saw my first one in that fully developed garb, it was during a Canadian summer and it stood out on the beach at Gaspé like a piece of rounded house-brick.

(One of my typos created the Red Not, which looks odd at first, until the penny drops. The Red Not was obviously a calidrid spotted near Murmansk during the Cold War. The Soviets denied its existence, of course.)

Other individual species names among our small sandpipers are generally straightforward. The specific name of the Sanderling, *Calidris alba*, refers to the whiteness of its plumage, particularly in the winter. A small bird that scuttled around on sand was bound to become a 'Sander' and then, being little, to gain a diminutive suffix and emerge as Sanderling.

Curlew Sandpiper, *Calidris ferruginea*, offers its 'rusty-red' breeding colour in the specific name, while the word Curlew simply reminds us to look out for a bird which resembles a miniature Curlew, in the bill-shape at least.

The specific name of Purple Sandpiper, *Calidris maritima*, confirms its adherence to sea coasts, though the colour is really only a purplish-grey.

The word 'stint' is thought to originate from an East Anglian name for Dunlin, and therefore the addition of 'Little' to its small relative makes good sense. Little Stint is only *Calidris minuta* (little), but among the vagrant species the rare Long-toed Stint is *C. subminuta* (somewhat small), Least Sandpiper is *C. minutilla* (very little), and Semipalmated Sandpiper is *C. pusilla* (very, very little). Who says that scientists have no imagination? But we can't say that they didn't warn us, as anyone who has struggled to identify these tiniest of waders will readily admit. And, since the gravely endangered Asian species, Spoon-billed Sandpiper, *Eurynorhynchus pygmeus*, has received so much publicity in recent years, it is worth noting that *pygmeus* is yet another way of expressing small size among waders. Its generic name means literally 'broad-billed'. Temminck's Stint is spared the 'shrinking game', since *Calidris temminckii* is named after C. J. Temminck, a nineteenth-century Danish ornithologist.

A number of other vagrant sandpipers appear in our guides – and hopefully on our lists too.

The Broad-billed Sandpiper, *Limicola falcinellus*, is a surprisingly scarce vagrant for a Scandinavian breeder. The English name describes a species which, though marginally smaller than Dunlin, has a proportionately solid bill. The scientific name tells us that is a 'mud-dweller with a sickle-shaped bill' (see Falcated Duck, Glossy Ibis and Falcon for other uses of the sickle image).

Much more frequent is Pectoral Sandpiper, *Calidris melanotos*, which has a specific name meaning 'black-backed'. However, its English name concentrates on the front, in the form of the striking pectoral (breast) shield of plumage worn by the breeding male. Neither of those is a great help in the identification of passage birds, though there is some subtle clue in the pattern of breast markings.

The very similar Sharp-tailed Sandpiper, *Calidris acuminata*, has names which match in English and Latin, while White-rumped Sandpiper, *C. fuscicollis*, has one distinctive feature in its English name and a totally different one in the name *fuscicollis*, which refers to the 'dusky neck' of the bird's winter plumage – both useful features in the field.

The long-legged Stilt Sandpiper, *Calidris himantopus*, takes both its specific names from the generic name of the stilts.

Baird's Sandpiper, *Calidris bairdii*, is named after S. F. Baird, a nineteenth-century US ornithologist, while Western Sandpiper, *C. mauri*, commemorates a nineteenth-century Italian botanist, Ernesto Mauri. The western bias also applies in America, in the bird's range there.

Voices in the wilderness

Of the word 'sandpiper', it is a most obvious thing to state that the word derives from the piping voices of birds that live on the sand. Colloquially the Americans call the small ones 'peeps' for similar reasons. However, other wader names derive from the sounds that the birds make and some are far from obvious. These include the Whimbrel (from the Northern English dialect name *whimmerel*); and the Curlew (via the French, *courlieu*), and some authorities suggest that the Anglo-Saxon Godwit may also be echoic (but more of that later.)

In America, the strange name Killdeer has no sinister tale to tell: it is merely imitative of the call of *Charadrius vociferus*. The specific name echoes the idea that it is noisy. Among the American waders, Willet, *Tringa semipalmata*, has a vernacular name from the same tradition, while Tattler gave the alarm for so many hunters that they simply accused it of 'tattling', as in telling tales. I have watched many Willets of both races while in Florida, and have seen Killdeer there, but my one encounter with a tattler species was in the form of the Grey-tailed Tattler, *Tringa brevipes*, which I found on the beaches at Darwin. The American species is Wandering Tattler, *Tringa incana* (light grey).

The long and the short of it... dowitchers

Killdeer is on the British list as a vagrant, as are the two dowitchers. Here there is a bit more colour in their story: Choate suggests that the name 'dowitcher' is a corruption of 'Deutscher' (German) or 'Duitsch' (Dutch) snipe, but that may well have been a European settler's approximation for a word which is thought otherwise to derive from an Iroquois form, or from the Mohawk *tawistawis* (snipe). These birds share many characteristics with the snipes, but unlike the snipes they have a distinctive reddish breeding plumage which is quite different from their grey-toned winter plumage. I have never appreciated it more than when I watched Short-billed Dowitchers blending with the grey-red mud of Copequid Bay in Nova Scotia, while Willets were more obvious in a brown-grey plumage. In Coward's 1926 list the bird was known as the Red-breasted Snipe, though in the 1923 text of his second volume he made mention of the American name in a secondary way.

The long or short bill of the two names is not a very useful guide to their identification, by the way, but I was safe in Nova Scotia in summer, well outside the range of Long-billed. The generic name is *Limnodromus*, from the Greek *limne* (marsh) and *dromos* (racer). In the case of the Long-billed, the specific name is *scolopaceus*, a reference to *Scolopax*, the name of the Woodcock and an older name of the Snipe. Short-billed is *L. griseus*, which means 'grey'. Since both are grey in winter, and since both were wintering at Sanibel Island in Florida, I found that their separation was not that easy.

Cryptic in plumage and in name: Woodcock and snipe

Woodcock, *Scolopax rusticola*, is an example of a name where the sexes count. It implies that it was one half of a duo woodcock/woodhen (as in peacock/peahen). Its scientific name *Scolopax* is Latin for woodcock or snipe, while the specific name *rusticola* means 'country-dweller' and may originally have referred to another game bird such as a grouse. The Common Snipe, *Gallinago gallinago*, links the idea back to the concept of the sexes in 'resembling a chicken' (Latin *gallina*). The vernacular name is more interesting, since it is derived from a Germanic source which refers to a long sharp instrument and thus to the bird's bill. It may be related to the word 'snip' as with scissors. Both Woodcock and Snipe have long been treated as game birds, but the Snipe is renowned as a difficult target: consequently the word is transferred to the skill of the sharpshooter or 'sniper'. The difficulty of shooting a snipe relates to a useful field-skill for the birder too: the Common Snipe zigzags on take-off and flies into the distance, while Jack Snipe, *Lymnocryptes minimus*, flies straight and drops back to the ground quite soon.

COMMON SNIPE

The generic name of this last species harps back to that of the Dowitchers, since the Greek word *limne* (marsh) also occurs in *Lymnocryptes*, with the second element taken from *krupto* (conceal). As birdwatchers know, this species is most difficult to spot when still: all the snipes have plumage designed to blend with their habitat, and the Jack Snipe is the master – until it moves. It is astonishing how the bird's bobbing horizontal cream stripes work against the verticals of the reeds to allow one to follow its progress. As *L. minimus*, it is the smallest of the snipes and in that case the name 'Jack' is used literally, in its old sense of 'small', which is discussed later in the name of Jackdaw.

Historically, the Common Snipe was often referred to as the Full Snipe, a form which contrasted it with its smaller cousin, while the term Summer Snipe was sometimes used for Common Sandpiper.

A good understanding: shanks

That dubious pun was one used by my late father to describe his growing son's big feet. Feet and legs do, however, play quite a part in wader names, and not least in the 'shanks' and their allies.

Legs are 'shanks', as in the nickname of Edward I, 'Longshanks', and in 'Shanks's pony', an ironic term for walking. To birders, shanks are birds of the genus *Tringa,* and the leg colour is important in identifying Greenshank, Redshank and the yellowlegs (Greater and Lesser). It is a great pity that the Old World traditions were not retained in the names of the two New World species, since Coward's 1926 BOU list had them as Greater Yellowshank and Yellowshank, which seems to work so well.

In the case of the Spotted Redshank, *Tringa erythropus,* the Greek-derived specific name describes the red feet (and legs). It is worth pausing here to remember that what we tend to think of as the 'knee' of a bird is the equivalent of our ankle, so a bird has a lot more 'foot' than we realise. Latin is used in *Tringa flavipes* to describe the yellow feet of Lesser Yellowlegs. However that second idea is also conveyed in the Greek, *Tringa ochropus,* but most unhelpfully it is the name of the Green Sandpiper, which normally has greyish-green legs. This appears to be down to Gesner in 1555, but was perpetuated by Linnaeus, in spite of his descriptive note including the phrase *pedibus virescentibus,* which stated clearly that the bird had greenish legs. It is entirely possible that Gesner had confused Green and Wood Sandpiper, which are broadly similar. However, a few years ago I did see an individual Green Sandpiper with distinctly yellow legs. On that occasion, I photographed the bird, scoured all available references, and consulted local experts, just in case I had found a vagrant juvenile Solitary Sandpiper, which may sometimes have yellow*ish* legs. Was a single aberrant specimen behind this odd choice of name? It seems highly improbable, since I have never seen another even vaguely yellow in the legs.

GREENSHANK

The Grey-tailed Tattler gets back into the frame with the name *Heteroscelus brevipes*, which states in Greek, in the first part, that the bones are uneven and, in the second, in Latin, that the foot is short. It all seems very precise information. Fortunately many authorities now prefer *Tringa brevipes*, which is a lot easier to spell.

As for the generic name *Tringa*, this seems for once to be clearly attributable to Aristotle's description of 'a thrush-sized, white-rumped wading bird that bobs its tail'. *Trungas* is Greek for thrush, and the present name was adopted by Aldrovandus in 1599, with reference to the Green Sandpiper, which seems to fit the bill nicely.

Legs do not, however, feature in the scientific names of several members of this genus. Wood Sandpiper, *Tringa glareola*, is named after a 'gravel habitat', while Marsh Sandpiper, *T. stagnatilis*, tells of its preference for 'marshy pools'. The aforementioned Solitary Sandpiper is *T. solitaria*, which rather speaks for itself (though I have to say that we saw two quite close together in Trinidad). Common Redshank is *T. totanus*, from an Italian name, *Tõtano*, for the species. Greenshank is *T. nebularia*, a name which concentrates on the 'cloudy grey' of the plumage.

The allies of the shanks bring two new generic names.

Common Sandpiper, *Actitis hypoleucos*, and its American congener, Spotted Sandpiper, *A. macularius*, take their generic name from a Greek word meaning 'coast-dweller'. Their specific names mean 'white undersides' and 'spotted' respectively, though this is only applicable to the latter in summer plumage. Wintering Spotted Sandpipers were a regular feature of the Tampa Bay waterside: apart from their yellower legs, they are broadly very similar in winter to Common Sandpiper. They were easy in a Canadian summer, when spotted like a thrush.

Another 'odd man out' is the American Buff-breasted Sandpiper, *Tryngites subruficollis*. The nineteenth-century German naturalist, Cabanis, is responsible for the creation of the generic name, which he linked to *Tringa*. The specific name means 'reddish-necked', though the pink-buff wash of the neck and breast is not well served by that description

The unconsidered trifles

In Shakespeare's *Winter's Tale*, Autolycus, the light-fingered peddler, is 'a snapper-up of unconsidered trifles'. This section deals with a miscellany of left-over items, starting with the most Autolycus-like of the waders.

The vernacular name of the Turnstone, *Arenaria interpres*, is self-evident to anyone who has watched the bird flip over stones and weed to find food on the beach. *Arenaria* means 'of the sand', but *interpres*, meaning 'messenger', is apparently just another of Linnaeus's misunderstandings. It appears that he was confused by a Gotland dialect word for Common Redshank and a Swedish word meaning 'translator/interpreter'. However, the idea of the 'messenger of the sand' scuttling about on errands is not a bad one.

A possible echoic origin of the word 'godwit' is mentioned above (though that would be a fanciful rendition of the call of either of the British species), so it seems more likely that the Anglo-Saxon root is *gód* (good), *wiht* (creature), as stated by both Webster's and Chamber's dictionaries. This idea might be supported by their culinary status, as cited by Muffet's quotation of their market price, and by the

OED's evidence of Sir Thomas Browne who referred to 'Godwyts... accounted the daintiest dish in England; and, I think, for the bigness, of the biggest price.' An alternative suggestion exists that the word relates literally to God and to the Anglo-Saxon *witten* (to know). That leads us God-knows-where and seems to be a clutched straw too many.

The generic name *Limosa* means 'muddy' (see *Limicola*, above). Black-tailed Godwit, *Limosa limosa*, does deserve to be muddy in stereo, because it frequently buries its head deep in liquid mud to retrieve its food. Bar-tailed Godwit, *L. lapponica*, always seems a less-messy feeder, since it tends to feed more on the tideline and by probing less deeply. The name *lapponica* (Lapland) underlines the fact that this is a more northerly species – and one which undertakes some marathon migrations. An American species, Hudsonian Godwit, *L. haemastica*, also appears on the British list. Named after Hudson's Bay in English, its scientific name tells that it is 'bloody', a feature which reflects the fact of its rufous-brown breast in summer plumage, which is several shades darker than the two British species. Curiously Marbled Godwit, which has not been recorded in Britain, bears the scientific name *L. fedoa*, which is thought to be an old English name, now of lost origins. Its earliest evidence was in Turner in 1544 and it was used by Edwards in 1750. The vernacular name is a description of the fine black patterns of the summer plumage on a pale, tawny-buff ground. Those which I saw at Weedon Island, Florida, in winter still had the warm colour, but had plainer undersides.

BLACK-TAILED GODWIT

Ruff, *Philomachus pugnax*, has a curious history, and appears to have been adopted as a name after the late Tudor/early Stuart fashion for wearing an exaggerated lace collar, known as a ruff. Lockwood points out that the name was first noted in 1634 in '12 Ruff and reeve 3 dozen'. The use of the word Reeve for the female may in fact be the older name for the species: Lockwood speculates that the original form was Ree, and that it fell into line with the use of the word

Reeve (a local officer). I feel that there is great significance in the evidence of the quoted market bill and that the explanation for the use of the two words lies in both commerce and sexual dimorphism. This species is unique among waders in having a male which is considerably larger and heavier than the female on a ratio of roughly 5:3 (BWPi data). That would clearly affect their relative market value as food items. The size of the male has evolved, as has its elaborate plumage, as part of its competitive lekking display, its aggression being doubly recorded in the scientific name *Philomachus pugnax* (see earlier comment). Given the bird's unique characteristics, I am surprised to see that the BOU has recently placed it in the genus *Calidris*, as *Calidris pugnax*, alongside such as Dunlin. Conversely, the most recent IOC list retains the traditional name and has not flagged any possible change.

Sitting close to the top of my 'most-wanted' list, the Terek Sandpiper, *Xenus cinereus*, has an apt name, since *Xenus* derives from the Greek *xenos*, meaning 'stranger'. The fact that it is grey (*cinereus*) doesn't make it any less interesting. Unusually, the name *Terek* is that of a river, in this case one rising in Georgia. Catherine the Great of Russia sent an exploratory and scientific expedition into the area in 1768 and one of its members was the German, Johann Güldenstädt, who named the bird in 1775. The same year he also described and named the redstart species which bears his name in the English form.

Closer in relationship to the curlews, Upland Sandpiper, *Bartramia longicauda*, was known to Coward's generation as Bartram's Sandpiper. William Bartram (1739–1823) was not only long-lived, but he was considered the grandfather of American ornithology, having supported Alexander Wilson (see Wilson's Plover), who is considered its father. The bird's long-tailed appearance accounts for its specific name, while the vernacular name points to the fact that it breeds in dry, open plains of North America and often at altitudes of 1,000 metres or more.

And a few more leg issues...

A most unusual observation is enshrined in *Numenius phaeopus*, the name of the Whimbrel, since the specific name means a 'dusky-brown foot', a description which could apply to an awful lot of waders and seems quite unhelpful in any case.

Semipalmated Sandpiper and the similarly named plover both have partial webbing between their toes to earn that name, which also occurs in the species name of the Willet in the Latin form *semipalmata*.

The word 'phalarope' was discussed earlier, but it seems appropriate to recall here that the word means 'coot-footed', because of the similarity of the bird's foot to that of the Common Coot. The Grey Phalarope, *Phalaropus fulicarius*, contains the idea in its specific name too, but that of the Red-necked Phalarope, *P. lobatus*, simply speaks of the shape of the foot as 'lobate', i.e. with lobes of skin. The name of the third species, Wilson's Phalarope, *P. tricolor*, refers (somewhat inaccurately) to the breeding plumage, which is grey, black, orange and white.

The curious name Thick-knee, which is often substituted for Stone-curlew, is approximated in the scientific name of our own species, *Burhinus oedicnemus*. With *Burhinus* meaning 'ox-nose' (a reference to the heavy bill), and *oedicnemus* meaning 'swollen-legged', this has to be one of the oddest of all scientific names. But Thick-knee is a very loose term to describe what equates to our 'heel' and seems a very odd choice. The name Stone-curlew is not particularly accurate either, since this species is not closely related to the curlews: only the size and call give it a claim to that name. Historically its names included Great Plover and Norfolk Plover, since its proportions at least are akin to those of such as Grey/Black-bellied Plover. This now-scarce summer migrant still has a stronghold in the Norfolk and Suffolk Brecklands, and still hangs on in the Wessex downlands. It is a crepuscular and difficult species, yet its cousin, the Bush Stone-curlew of Australia, proved much easier to see well, particularly in the Northern Territory.

BUSH STONE-CURLEW

And, before I leave waders once and for all, another bird for the typo 'listers': the male Bra-tailed Godwit has a red face in all plumages.

LEADING THE FIELD: COURSERS AND PRATINCOLES

The coursers and pratincoles are occasionally represented by vagrant birds to Britain. They have long proved problematic for the taxonomists, since they seem to have some affinity with the waders and some with the terns. Their seventeen species form the Glareolidae, a word which is based on the Latin for 'gravel'.

The Collared Pratincole is *Glareola pratincola*. In this case the specific name also refers to habitat, since it means 'meadow-dweller', as does the Anglicised form. In fact we encountered the species twice in grassy habitat in Andalucia. The Black-winged Pratincole, *Glareola nordmanni*, takes its name from the nineteenth-century Finnish naturalist Alexander von Nordmann, whose name also appears in Nordmann's Greenshank.

The coursers are runners, their name derived from the Latin *currere*, to run. Cream-coloured Courser is *Cursorius cursor*, but in spite of that I think it comes in second to the Spotted Pardalote for repetition of a good idea.

SNATCHERS KEEPERS: SKUAS AND JAEGERS

There is little comfort, I feel, in the issue of names in the family Stercorariidae. At least the second edition of the *Collins Bird Guide* abandoned the horror of Parasitic Skua, which it had used in the first edition in 1999. That had annoyed me almost as much as the abominable concoction Great Northern Loon. It has now been restored to Arctic Skua, a form which British birders find a lot more comfortable. The accepted name in North America is Parasitic Jaeger. The scientific name is *Stercorarius parasiticus.*

Let me deal with 'skua' first. The word originates in the Faroe Islands as *skúvur* and travels via Norway after 1604 as *sku*, to be Latinised subsequently in the form *skua*. Ray Anglicised the word in 1678. The American preference for 'jaeger' is derived from the German for hunter.

As for the 'parasitic' element of the name, it is explained in part by jaeger. Strictly speaking the bird is a kleptoparasite, in that it steals from other birds (mainly terns) by pursuing them until they drop or regurgitate their catch. The consumption of regurgitated food leads to the scientific name *Stercorarius*, which literally means 'dung-eating'. The fact that it cohabits mainly with Arctic Terns may have influenced the choice of the English common name, though I have watched a pair hammering a flock of Sandwich Terns off Dawlish Warren. For an Arctic Skua, anything goes.

Sailors once used the name Marlinspike Bird, a reference to the extended central tail spike. The marlin spike is a pointed tool used by mariners to separate the strands of rope during splicing (or 'marling'). It is hardly surprising that the language of seamen gets into the frame: frigatebird is another such form, which is still used to name a family of tropical birds with similar habits. Having had an up-close encounter with a perched Great Frigatebird, *Fregata minor*, on the Great Barrier Reef, I can vouch for the fact that its bill looks rather like a naval cutlass.

The Long-tailed Jaeger/Skua, *Stercorarius longicauda*, offers us no colour in its specific name, other than the fact that it echoes that of the Upland Sandpiper. (The typo form, the Log-tailed Skua, finds it harder to fly, but floats better.)

I mentioned Bonxie earlier, but a Cornish name for the Great Skua, *Stercorarius skua*, was Tom Harry. Such use of personal names was quite normal, but the spelling variants which occurred in Harry suggested to Lockwood that the word may have had an older, lost origin, particularly as it appears in the verb 'to harry' - which the Great Skua certainly does. This verb also provided the origin of the word harrier among the raptors.

The Americans do as they do with bird names, but I was long puzzled by use of the word skua for the larger species and jaeger for the two smaller ones. There has also long been hesitation about where Pomarine Skua fitted. The distinction is rooted in a former division between *Catharacta* (rapacious) and *Stercorarius* (parasitic) skuas. Great Skua and its southern allies are considered to be primarily predatory in their behaviour, hence the distinction, with Pomarine being partially so. Modern DNA studies have shown that all the skuas are more closely related than was thought, and the former *Catharacta* skuas have now been lumped in with the *Stercorarius* species. In spite of that move, the IOC list's formal version of the English names now confirms that the correct forms are Great Skua and

Pomarine Skua, but Parasitic Jaeger and Long-tailed Jaeger. In this case, there was apparently no use of the rule of precedence which occurs in scientific names. The name Arctic Skua was created by Fleming in 1828 off the back of an older, if erroneous, name, Arctic Gull, used by Pennant. The OED tells us that the word 'jaeger' was first applied to the skua in 1838. (With that and the Common Loon, any residual British smugness over the Long-tailed Duck should now have vanished.)

In truth this seems to be an area of historical turbulence. Hartert's 1912 *Hand-List* uses the names we know today, though with a strange spelling of Pomatorhine. However, in 1926, Coward used two unfamiliar names: for him, Arctic Skua was Richardson's Skua and Long-tailed Skua was Buffon's Skua. So before we get huffy about change, we must remember that such turbulence buffeted other generations too.

And, while I am here, Pomarine and *pomarinus* (as in *Stercorarius pomarinus*) both derive from the Greek *pōma, pōmatos* (lid, cover) and *rhis, rhinos* (nostrils) – which sources are much closer to the spelling used by Hartert and Coward. However, Lockwood points out that the current form is the older one, first derived from Temminck's *pomarinus*, an inaccurate adaptation of the Greek. The form Pomarine was used by Stevens in 1826, but eventually replaced by the BOU in 1883 in favour of the 'corrected' version Pomatorhine. Since that form hardly trips off the tongue, expediency has now supported historical precedence by reinstating the original form. Max Nicholson proposed a formal change in 1949, though the pedantic version still appeared in the 1952 *Checklist*. Since that date it has been dropped in favour of the simpler form. (Incidentally, Nicholson also prevailed in getting Dunnock reinstated in preference to the inaccurate Hedge Sparrow.)

But why was the strange-looking word created in the first place? It actually refers to the cere, a waxy membrane at the base of the bill of certain species. Why this feature, which is present in all skuas, was significant to Temminck in the naming of this species is not clear, especially since the word is not used for any species with more obvious ceres, such as those in the parrot, raptor and pigeon families. It is a pity that the ancient Baltic state of Pomerania, whose very name means 'the seashore state', doesn't get into the frame, but it seems fitting to say that it is a red herring.

LARIDOPHILE OR LARIDOPHOBE? GULLS

The word 'gull' derives from a Celtic root, coming into general English usage via Cornish in about 1430. For many centuries the term 'seagull' has been a common usage, since gulls are primarily coastal birds and some, like Kittiwake, are truly pelagic. However, in twentieth-century Britain many developed increasingly inland habits. This has as much to do with human modification of the sea, coastline and landscape as with the gulls, which are opportunists. In a chatty book written for children in 1909, Richard Kearton tells how the very harsh winter of 1895 drove huge flocks of Black-headed Gulls up the Thames, where they were fed by delighted Londoners, such was the novelty. With reference to gulls feeding on crop fields, he actually observes that 'one day we will have to change their name from Sea to Land Gulls'. Today, with many gulls spending much of their year inland, we would readily see the wisdom of that prediction, although the simple term 'gulls' seems to cover all eventualities perfectly well.

The generic name *Larus*, which is used for many gull species, is a Latin word rooted in the Greek *laros* (a gull). The word Laridae derives from that word and now embraces noddies, skimmers, gulls and terns. The gulls form one of the most difficult subsets of birding, because a number of similar species have several plumage phases and the largest take four years to mature, so they each take on a number of appearances. Gulls are like the game of cricket or like Marmite: they divide humanity into two camps. Those who are fascinated by the challenge or love their gulls tend to call themselves laridophiles or larophiles, so those who confess to total bafflement ('I don't do gulls!') must be laridophobes or larophobes. In truth there is more of a middle ground than with cricket and Marmite. I tend to find them fascinating, but confess to only modest skills. But this is not about the difficulty of identifying gulls: rather it is about the complex issues of some of the names, which is bound to make the laridophobes even more entrenched.

YELLOW-LEGGED GULL

Firstly there is the issue of the proliferating species, which has puzzled bird-watchers for some time. Once there were just Herring Gulls and now there are European and American Herring Gulls, Yellow-legged Gulls, Caspian Gulls, Armenian Gulls and others. This is the result of the careful consideration of distinct populations, of their distinct characteristics and behaviour, and of genetic studies, which together have resulted in 'splits' – the naming of new species within a former single species. Closer definition of some subspecies leads to such references as Scandinavian Herring Gulls and Baltic Lesser Black-backs. It is all a bit confusing to Joe or Jane Birder, but when I learned that Yellow-legged Gulls have a different moult timing which makes them stand out in the autumn crowd, and after I had seen their quieter behaviour in a breeding colony on Gibraltar, I became more convinced. I had always felt that the American Herring Gull was bulkier than the British one, having been on handshake terms with them as I watched Tampa Bay at dawn while they perched quietly a few metres away. Studies have now shown that this species descends from a different ancestral stock from those of Europe. I can now discern the different shape of the Caspian Gull, so I now forgive the birding magazines and the laridophile entries on the rare-bird alerts for the years when I was baffled by names which were not in my old handbooks.

To make things even more complicated, the scientific names of some gulls are often unhelpful, to say the least.

Let me start with the revised genus for the Black-headed Gull and its kin. The inelegant 'new' name, *Chroicocephalus*, is a reversion to a form created by Thomas Eyton (1809–1880) in 1836. It derives from the Greek *khroizō* (to colour or stain) and *cephalus* (head). Well, that makes moderate sense for a group that includes Black-headed, Brown-headed and Grey-headed Gulls, but seems quite inappropriate for the pristine, uncoloured heads of the ubiquitous Silver Gull in Australia, and of Europe's Slender-billed Gull. The name seems to be a clumsy mouthful and offers limited accuracy in its content.

The Little Gull has also been reclassified, but as the unique member of a 'new' genus, again one with a curious name, *Hydrocoloeus*. This derives from the Greek *hudro* (water) and *koloios* (a type of web-footed bird, possibly a cormorant). Since gulls are *all* web-footed water birds, this is hardly helpful. It is in fact a return to an 1829 name created by the German naturalist Johann Kaup.

I assume that some re-tuning of gull classification was needed to differentiate between gulls which were previously 'lumped' in *Larus*, and that those last two rather unlikeable names have to take priority under the rules of the Linnean system. That seems a pity, since there is a good precedent for creativity within this family. I recently saw a Sabine's Gull, *Xema sabini*, and was intrigued by the beauty of the bird and the monochrome geometry of the wing-patterns. But then I caught up with the oddity of the word, *Xema*, and was even more intrigued. So were others, it seems, since Jobling quotes several bemused commentators. The consensus is that it is a nonsense word, made up to do a job and no more. Whether it was the product of a tired and uninspired mind, a bit like the old Friday-afternoon washing machine, or whether Leach was having a little joke, is irrelevant: it is weird and it is unique, and strikes me as a lot better than names whose pretentions are rather pointless. Sabine's Gull, incidentally, commemorates General Sir Edward Sabine (1783–1883), an Arctic explorer and President of the Royal Society.

SABINE'S GULL

If the rationale behind those names was baffling, there is a lot more in the world of gulls to make the head whirl. In the scientific names, Black-headed Gull is the 'laughing' gull (*ridibundus*) and Caspian Gull the 'laughing aloud' gull (*cacchinans*). Mediterranean Gull is the 'black-headed' gull (*melanocephalus*), leaving Laughing Gull as merely the 'black-tailed' gull (*atricilla*). Since the scientific names form a universal baseline, any inconsistency must be in the vernacular forms. But that 'black tail' of the Laughing Gull is of course white. Linnaeus seems to have misread his own notes and mistakenly used *atricilla* when he intended a*tricapilla* (black-headed).

What is more, the Black-headed Gull, with its dark-brown cap, has long been wrongly named, though Brown-headed Gull was sometimes used in the nineteenth century. Today that common name applies to another bird of that genus, while a third one is Brown-hooded. Can you feel the Edvard Munch moment coming on? Even the much restricted use of the name Herring Gull still has some power to confuse, since it is *Larus argentatus*, while the subspecies which breeds in Britain is *L. argentatus argenteus*. Both names are variants of 'silver-coloured'.

Further oddities occur in the naming of some of the more northerly gulls. There is no difficulty with *Larus hyberboreus*, which tells that the Glaucous Gull is 'very northerly'. It takes its English name from the delicate blue-grey colour of the adult's mantle and wings. However, the smaller Iceland Gull is similar enough to be named *Larus glaucoides*, which means 'resembling glaucous'. Furthermore its vernacular name is somewhat misleading, because it breeds no further east than Greenland and, to add to our confusion, its Canadian race is popularly known as Kumlein's Gull.

There is much less confusion in the names of other gulls.

Great Black-backed Gull is *Larus marinus*, which clearly suggests that this is a gull of the sea-coasts. We saw its Australian equivalent in Tasmania, the impressive Pacific Gull, *L. pacificus*, which sports a broad bill almost worthy of a Puffin. On the same beaches were the smaller Kelp Gulls, *L. dominicanus*, which take their vernacular name from the giant seaweeds and their specific name from the resemblance of their pristine black and white to the garb of Dominican friars. These were the equivalent of the northern hemisphere's Lesser Black-backed Gull, which is *L. fuscus* (dusky), though the race we see in Britain is normally *L. f. graellsii*. That subspecific name recalls a nineteenth-century Spanish zoologist, who is truly worth a place in this context: his full name was Dr Mariano de la Paz Graells y de la Agüera.

Other eminent scientists feature frequently. Slender-billed Gull, *Chroicocephalus genei*, lives up to its English name, but the specific name *genei* commemorates Giuseppe Gené, a nineteenth-century Italian naturalist. The Yellow-legged Gull, *Larus michahellis*, is named for the nineteenth-century German zoologist Karl Michahelles. His contemporary, the French naturalist Jean Audouin, appears in the name of Audouin's Gull.

Formerly known as the Great Black-headed Gull, Pallas's Gull records the name of the German naturalist Peter Pallas (1741–1811), who appears in a dozen other species' names, mainly as a result of joining an expedition of the Academy of Sciences of St Petersburg to explore remote areas of Russia. Its scientific name, *Ichthyaetus ichthyaetus*, clearly labels this very large gull as formidable, since it means 'fish eagle', though the much more modest Mediterranean and Audouin's Gulls are also classified with it, at least in the IOC list.

MEDITERRANEAN GULL (WINTER)

Among the American vagrants, Bonaparte's Gull, *Chroicocephalus philadelphia*, commemorates Charles Lucien, a nephew of the Emperor, who became an eminent ornithologist while living in exile in the US (there is more about him later). Here the specific name *philadelphia* is derived from the city in Pennsylvania. The American Herring Gull, *Larus smithsonianus*, takes its name from James Smithson, 1765–1829, mineralogist and chemist, whose bequests founded the Smithsonian Museum.

Franklin's Gull is dedicated to Sir John Franklin, 1786–1847, the explorer of the Northwest Passage. However, this bird bears an intriguing scientific name, *Leucophaeus pipixcan*. While the generic segment, which is shared with Laughing Gull, simply means 'ash-coloured', the specific *pipixcan* comes from the much warmer climes of Mexico and was apparently a local Amerindian name. Relatively few names of this sort make it into the early Linnean records, but this one comes from the explorer-naturalist Francisco Hernandez de Toledo (1514–1587), whose drawing of it in the records was considered to be an acceptable representation of the species which winters that far south. It was so named by the German, Johann Wagler in 1831.

Ross's Gull takes its name from the Arctic explorer Rear Admiral Sir James Ross (1800–1862) who was also commemorated in the name of a South Polar sea. The gull is the sole representative of its genus. Its name *Rhodostethia rosea* tells us in Greek that it is 'rosy-breasted' – and then underlines the rosiness in Latin. (A different Ross, Bernard, a chief factor at the Hudson's Bay Company, is attached to Ross's Goose.)

Clearly most northerly of all, Ivory Gull is *Pagophila eburnea*, which means 'ivory-coloured sea-ice lover'. Constantine Phipps, a naval officer, who became Lord Mulgrave in 1775, originally named and described the bird in 1774, but the generic name was a later invention of Johann Kaup in 1829. Between them they did an excellent job, since the name sums up the bird beautifully.

Ring-billed Gull, *Larus delawarensis*, commemorates the Delaware River of the US. The similar Common Gull is officially the Mew Gull on the IOC list, the American name having prevailed. In truth, the British form is not the most common of our gulls and, as *Larus canus* (grey gull), it has a dreary name. The North American Mew Gull is the subspecies *brachyrhynchus*, which name tells us that it has a shorter bill.

One of my great favourites, both as a species and as a name, has always been the Kittiwake, *Rissa tridactyla*. Anyone who has heard the call will immediately know the origin of the name Kittiwake, since the bird calls its name constantly near the cliffs on which it breeds. However, it is strictly the Black-legged Kittiwake, since there is a Red-legged species, *R. brevirostris* (short-billed), on the Pacific coast of North America. The generic name *Rissa* is a corrupted version of the Icelandic name, *Rita*, for the species. The specific name *tridactyla* means 'three-toed' and relates to the fact that it has no rear toe. Curiously, a Pacific subspecies, *pollicaris*, is so called because it has a partly developed *pollex* (thumb). All that said, the Kittiwake is a beautiful bird.

And in summary, two things are fairly certain: the first is that gull-naming is still far from fixed and final; and the second is that whatever gulls are called by name, they will never be called *easy* by the gull-watchers.

ONE GOOD TERN DESERVES A SEA SWALLOW

The graceful Sternidae, once a family to themselves, are currently classified within the Laridae. The generic word *Sterna* is not, as might appear, a word of Latin origin. In fact it is English, based on *stern* (variants *stearn, starn*), which Turner Latinised in 1544. A related form tern (or *tarn*) was noted by Ray in 1678, but he preferred Sea Swallow. Once more, it was Pennant who confirmed the word 'tern' in 1768.

It is the Common Tern, *Sterna hirundo*, which formally echoes Ray's preference, with *hirundo* (swallow). The name Arctic Tern reflects the more northerly range of the species and is no surprise, unlike the name *Sterna paradisaea*. The simple explanation for that name lies in a feature which all birdwatchers look out for – the long tail streamers which project beyond the wing-tips of the perched bird. It occurs in the name of the Buff-breasted Paradise Kingfisher, a species which we saw in the rainforests of Queensland, its magnificent white tail-plumes its outstanding feature. The link between the tail and paradise is simply the concept of exotic beauty.

During the breeding season, Roseate Tern, *Sterna dougallii*, develops a rose-coloured tinge to its breast feathers to earn its common name. Dr Peter MacDougall is commemorated (somewhat inaccurately) in the specific name. He sent a type specimen to Montagu from the Firth of Clyde in 1813.

The North American Forster's Tern, *Sterna forsteri*, takes its name from J. R. Forster, the naturalist who replaced Joseph Banks on Cook's second Pacific voyage. In winter its distinct 'Zorro' mask makes it easy to identify, but in summer it is confusingly similar to Common and Arctic Terns.

And that brings me to the 'Commic' Tern, the fence-sitters' name for the difficult mobile bird which is hard to identify positively. Whoever first coined that phrase was onto a winner. My most confusing tern-watch was on one April morning which produced hundreds of terns of some eight species over Tampa

Bay, with Common, Forster's and even a few Roseate Terns all mixed in with Gull-billed, Royal, Caspian, Cabot's and Least Terns. To say that it was one of the most challenging mornings would be no exaggeration.

Of our five breeding species, the smallest is Little Tern, *Sternula albifrons*, the generic name being a diminutive of *Sterna*, and the specific name meaning 'white-fronted', a feature which is unique in the breeding plumages of the British species. Its congener, the Least Tern, *S. antillarum*, became very familiar in Florida, but its specific name records the Caribbean Antilles. We saw another close relative, the Yellow-billed Tern, *S. superciliaris* (meaning 'eyebrowed'), in Trinidad, while Australia gave us several sightings of Fairy Tern, *S. nereis*, which is named after the Greek sea-nymphs, the Nereids, fairies of a different sort.

The origin of the common name of Sandwich Tern, *Thalasseus sandvicensis*, was discussed earlier. Recent studies have now placed the species in the genus *Thalasseus*, from the Greek for 'fisherman'. The American form, Cabot's Tern, is now classed by some as a full species, *T. acuflavidus*. That name reflects a 'yellow needle' of a bill, which is only accurate for the subspecies known as the Cayenne Tern. The common name commemorates Samuel Cabot, a nineteenth-century American naturalist. Royal Tern gets its name from being the biggest of the genus, as *T. maximus* confirms. Both American species were frequent in Florida, and we also saw many Royals in Trinidad. The nineteenth-century German naturalist C. H. Bergius appears in the name of *T. bergii* (Greater Crested Tern), which we saw breeding by the thousand on the Great Barrier Reef, together with a colony of *T. cristata* (Lesser Crested Tern).

SANDWICH TERN

On that latter visit we also saw terns of a wholly different genus, *Onychoprion*. That strange word means 'saw-clawed' and refers to the fact that their feet have indented webs which make the claws stand out like saw-teeth. Ours were Sooty Tern, *O. fuscatus*, whose specific name means the same as the English. In Florida, one February, I saw a Bridled Tern, *O. anaethetus*, which means 'stupid'. This, like *Morus* and booby, is related to its trusting behaviour while nesting. 'Bridled' refers to the dark facial patterns, which recall the bridle worn by a horse. One Caribbean name for this highly pelagic species is Hurricane Bird, since it is only driven inland by storms: significantly, my bird appeared during a very strong southwesterly wind which told of storms to the south.

Our trip to Michaelmas Cay on the Great Barrier Reef also brought us into close contact with large numbers of another 'stupid' genus. These were Brown Noddy, *Anous stolidus*, and a few of the much smaller Black Noddy, *A. minutus*. The curious name noddy has a slightly obscure meaning, but it too relates to the fact that nesting birds appeared to be stupid and unafraid: the old slang word and nickname, 'Noddy', meant a stupid person, i.e. with a nodding head. The scientific names of the Brown Noddy mean 'unmindful' and 'impassive' respectively and underline that concept. The status of these very tern-like birds has recently been reviewed, and they now appear in the IOC list in Laridae, but separated from the terns by the whole block of gulls. But it is time to return to some European species...

In spite of the fact that this is a large, gull-like tern, the concept of 'water swallow' reappears in the generic form of the Caspian Tern, *Hydroprogne caspia* (Caspian water-swallow). Here again is the legend of Procne, Philomena and the enraged King Tereus (see Lapwing and Swallow). The reason for its attachment to the Caspian Sea lies in the fact that it was described and named by the German naturalist, Pallas, in 1770, during a six-year expedition in the area, while he was working under the patronage of Catherine the Great. The bird has a far wider range, of course: it is a vagrant to Britain, but I saw my first one on the Mediterranean coast of France, and many more in Florida.

Three species of another genus are seen in this country. The word *Chlidonias* also echoes the swallow analogy: it is based on the Greek for 'swallow-like'. The term 'marsh tern' (as opposed to 'sea tern') describes a group which prefers inland areas of water and marsh. All three are darker, as implied in the name of Black Tern, *Chlidonias niger*, and White-winged Black Tern, *C. leocopterus* (Greek, *leukos*, white, and *pteron*, wing). The third, Whiskered Tern, *C. hybrida*, gets its English name from a white area of plumage on the side of the face, which stands out from the dark grey of the breeding plumage. The odd scientific name presumably arises from the fact of its intermediate coloration between sea terns and the other marsh terns.

The Gull-billed Tern, as the name implies, has a gull-like profile. In *Gelochelidon nilotica*, the generic word means 'laughing swallow' (the first element relating to the rapid alarm call), while the specific name means 'of the River Nile'. It is one of the most widespread of terns, which I have encountered in three subspecies in Florida, Australia and Spain. In habits this bird appears to fall between marsh and sea terns, since I have seen it on beaches on the Gulf of Mexico and in the billabongs of Kakadu.

The oddest of all the terns which I have seen was, without doubt, the Large-billed Tern, *Phaetusa simplex*, a Trinidad sighting. In mythology, Phaetusa personified radiance as a daughter of Helios, god of the sun, while *simplex* means 'plain'. That is a paradoxical name if ever there was one. The German naturalist Johann Freidrich Gmelin (1748–1804) was responsible for that oddity in 1789 (perhaps he had one ear on the news of the storming of the Bastille). The tern exhibits radiance in the form of a huge, slightly decurved, yellow bill which would have tempted me towards the name *Musarhynchos geestii* (that is today's homework: work it out).

Putting such facetiousness to one side, I have to say that terns have given me some of my most wonderful birding experiences: views of thousands on the

Great Barrier Reef; an Arctic Tern playing woodpecker on my head on Inner Farne; and the eight species fishing over Tampa Bay in a single morning. Yet the Little Tern will always remain my favourite, as the one I knew best when young – a true sea-swallow.

LITTLE TERN

THE AWKWARD SQUAD: AUKS

The complications and anomalies which emerge in the names of birds within this family readily remind me of the military term for the most uncoordinated recruits.

One traditional English name which took root in the New World was the Cornish word *murre*, which was originally noted by Ray in 1678. The Welsh form *morr* suggests a common Celtic origin. Originally it meant Razorbill, but became attached to Guillemot too. In North America it is now used as the name for a number of auks and provides another of those areas of birding where you have to keep on your toes because of the language difference: for example Brünnich's Guillemot is a Thick-billed Murre in North America (and now on the IOC list). Given its older, British roots, we Brits should not be troubled by the fact that murre has now been preferred on the IOC list, though it will be hard to break the habits of several hundred years. The truth is, in fact, that the word Guillemot, which we might be tempted to defend, was imported from French by Ray in 1678. Its diminutive ending had the sense of a young bird. It seems likely that it was a bit of an affectation when adopted, as something of a cut above dialect words, but Pennant later used it too, and in a generic way, to link Common Guillemot with Black Guillemot. Lockwood argues that the origins of the word seem to be in the most common personal name, Guillaume (William), and that the Scottish and Sussex names for the species, such as Willock, Wilkie and Will are not coincidental. (From now on, I will see the smartly uniformed ranks on the cliff-ledges as the Old Bill.)

COMMON GUILLEMOTS

It is interesting to note that, in 1926, Coward's BOU list distinguished between Northern Guillemot and Southern Guillemot, each with subspecies status. Though we tend not to use these names now, the former is the nominate race, *Uria aalge aalge*, while the southern form is *U. a. albionus* (Albion being an ancient name for Britain). The species had earlier been classified by Linnaeus as *troille*, a promising-looking word, with hints of a troll, but the mundane truth is that it was named after a Swedish Archbishop, Troilius. The bridled Guillemot, which occurs more in the north of the range, is simply a variant, rather than a subspecies or a species.

Not too surprisingly, the Scandinavian languages also feature strongly in this family. The word auk, the English name for the family to which these birds belong, originates in Old Norse *alka*, and is found in the Latinised *Alca* (as in *Alca torda*, Razorbill), and again in the scientific family name Alcidae. It also appears in the Danish form in *Uria aalge* (Common Guillemot). A visit by Linnaeus to the remote Swedish province of Gotland in 1741 results in the specific name *torda*, from a Gotland dialect word for Razorbill, while *grylle* (in *Cepphus grylle*, Black Guillemot) is another such word. The specific name *lomvia* (*Uria lomvia*, Brünnich's Guillemot) also has Scandinavian origins, as a word for auk or diver. It is *Polarlomvie* in modern Danish, the language and nationality of Morten Brünnich, a contemporary of Linnaeus and Pennant.

The two non-Scandinavian words in that last paragraph, *Uria* and *Cepphus*, are both of Greek origin and both mean 'a water bird'.

The Black Guillemot has also been able to hang on to a dialect name of Norse origin. James Fisher gave the word Tystie some modern validity in 1966 by pointing out that the bird was not strictly a guillemot. This attractive name, a reflection of the bird's call, is probably as well known today among the general birding public as the aforementioned word Bonxie. The bird was also known by Scottish seamen as the Dovekie, who likened the behaviour of a pair to that of doves: the addition of *–kie* was an affectionate form. That word travelled with emigrants to North

America, where it is now used for the Little Auk, *Alle alle*, though not for the Black Guillemot, which is known by the same name on both sides of the Atlantic. However, the Black Guillemot has a close relative on the Pacific shores of that continent which retains the link: it is named Pigeon Guillemot, *Cepphus columba*.

As for the Little Auk, its scientific name *Alle alle* is something of a careless accident, perpetrated by none other than Linnaeus himself. *Allē* is the Lapp name of the Long-tailed Duck and is yet another onomatopoeic rendering of the bird's call. It appears that Linnaeus may well have confused the small white sea-duck with descriptions of the winter plumage of the Little Auk. The result is that the Little Auk is more than confused on both sides of the Atlantic.

The vernacular name for the Razorbill needs little explanation, other than to point out that the razor in question is of the 'cut-throat' variety. This imagery may also be supported by an old Scottish name, Coulter, which derives from the Latin for knife, and is best known as a word for part of a plough. In like vein, Coulterneb (knife-nose) is an old Scottish name for the Puffin.

ATLANTIC PUFFIN

The Puffin, *Fratercula arctica*, has names which originate quite differently, in spite of the fact that it was classified by Linnaeus himself. The word puffin, which was created to describe the fat baby Manx Shearwaters, was generally later confused with the name of the auk and became attached permanently to it by the end of the nineteenth century. And, such is the oddity of language and usage, that a word which started life as the diminutive of a different species acquires further diminutive words for its young in the forms Puffling and Pufflet. Puffin is now a generic vernacular name which has safely crossed the Atlantic to include the western American species Tufted Puffin, *Fratercula cirrhata*, and Horned Puffin, *Fratercula corniculata*, as well as Atlantic Puffin, *Fratercula arctica*, which is found on both sides of the pond. Until recently, three subspecies of Atlantic Puffin were recognised, and these appear in some older references. However, the IOC list now notes that the species is considered to be monotypic and clinal. In simple terms, there are no subspecies and there is some gradual variation over the geographical range.

The word *Fratercula* is a Medieval Latin word meaning 'little friar', not even ornithologists being able to resist the anthropomorphic charm of an upright bird with a comical appearance which reminded them of a dumpy little friar in a black and white habit. While I am sure that a 1915 BOU comment about its gregarious habits was an unintentional pun, McCleod did write in 1954 that 'the bird rises from the sea, clasping its feet as though in prayer'. Was he related to the Northumbrian granny of the Long-tailed Duck, I wonder?

My own closest encounters with the auks were, in fact, on the Farne Islands of Northumberland, where we spent a day on two of the islands. The memory is of being almost knee-deep in birds, surely one of the best birding spectacles ever. In my memories it is right up in the same league as the billabongs of Kakadu, the nesting colonies on the Great Barrier Reef, the reserve on Sanibel Island, and the balconies of the Asa Wright Centre. And it's a lot closer to home.

ON SPLENDID WINGS: SANDGROUSE

The two sandgrouse species which inhabit Spain have so far eluded me, but they represent an odd family which has some characteristics of pigeons, waders and partridges.

The word 'sandgrouse' most helpfully points out that this is a grouse-like bird of arid regions and deserts. Neither Black-bellied nor Pin-tailed hold any mysteries, because both names point out characteristics which would be vital in the field.

The generic name is *Pterocles*, meaning 'splendid-winged'. Both European species do have strikingly zoned wings. In the case of the Black-bellied Sandgrouse, *P. orientalis*, the implication of an oriental bias in its distribution is misleading, though there are other populations further east. The specific name of Pin-tailed Sandgrouse, *P. alchata*, has a much more interesting origin, since it derives from the Arabic name for the species, which is *al kattar*, an onomatopoeic form, which appears as *kata* in Linnaeus's notes

WANING AND WAXING FORTUNES: DOVES AND PIGEONS

The species of dove and pigeon in Britain and Europe are limited in number, but they have some fascinating stories to tell.

Where turtles fly

I have witnessed many magical moments among waders on the shoreline, but nothing was more wonderful than standing within a few metres of a huge Leatherback Turtle which was laying its eggs on a Trinidad beach. But it is the feathered Turtle, 'the wedded turtle with hir herte trewe', of Chaucer's *Parliament of Fowls*, which now needs some consideration.

The Turtle Dove, *Streptopelia turtur*, appeared in that confusing guise for many centuries, as in Shakespeare's 1601 poem, 'The Phoenix and the Turtle'. Only once does the Bard use the word 'dove' in the poem, and the only clue is that it substitutes the word 'turtle'. In truth the confusion for us would not have been so for Shakespeare's contemporaries, or for earlier generations, since its use derived from the Latin *turtur*, which was almost certainly an onomatopoeic rendering of the purring call and is commemorated in the scientific name. The combined name, Turtle Dove, was first recorded in 1300, but was only taken up in ornithology by Turner in 1544. It clearly took some time to become the norm, since the archaic form also appeared in the King James Bible and lingered via that into modern times, although Turner's version was widely used by subsequent ornithologists.

Today the sound of the Turtle Dove will send shivers up my spine, so rare has it become in this country. Its decline has been dramatic and it seems to have failed to overcome disadvantages at three levels: it has suffered from agricultural practices in its summer range in Britain; from hunting on its migratory routes; and from habitat degradation and subsistence hunting in its African wintering areas. When I was young it was a common bird and the Collared Dove was unknown in Britain. Now the Collared Dove's dynamic expansion has made it a common species and the Turtle Dove could disappear in a puff of extinction, like the one in Shakespeare's poem. If only it could learn from its cousin. Or, more realistically, if only humankind as a whole could realise its responsibilities to the world it so readily exploits.

TURTLE DOVE

The eighteen bird...

The Eurasian Collared Dove, *Streptopelia decaocto*, is now a much more familiar bird and is closely related to the Turtle Dove. The word *Streptopelia* links the Greek word *streptos* (collar or neck-chain) to *pelia* (dove). Whereas the Turtle Dove wears a little black and white neck-patch, Collared Dove has a neat black stripe. The name 'Ring Dove' offers some potential for confusion, since it is an old name for the

Woodpigeon. It was still in use in Essex when I was young, sometimes corrupted to the form 'Ring Dow'. It has parallels in Dutch and German.

My first encounter with a bird resembling a Collared Dove was in an aviary in Paris in about 1953. It struck me as a most beautiful bird. On reflection it was almost certainly a Barbary Dove, a smaller species, and the domesticated version of the African Collared Dove. Little did I dream that within that very year birds just as beautiful would start to appear in gardens in Britain. Prior to its remarkable range expansion in the twentieth century, this species was restricted to southern Asia, though it had possibly penetrated southeastern Europe by 1838, when it was described and named by the aristocratic Hungarian scientist Frivaldszky. The specific name reflects the Greek myth of a maidservant paid just eighteen (*decaocto*) pieces. She bemoans her fate and is transmuted by the gods into a dove. Now that's what I call a good name! The three-note call of the bird is so distinctive and always reminds me of the joyous theme of Beethoven's *Pastoral Symphony*, rather than of any sort of complaint. Not that Beethoven would have been aware of the bird in 1811, but perhaps the existence of Nightingale, Quail and Cuckoo in the score help to reinforce this sound image in my mind. In more modern times the Eurasian Collared Dove has become artificially established in North America, where its spread has been dramatic.

...and the Emperor's niece

If any dove species deserves better to be associated with such a legend as the one above it is the Mourning Dove, *Zenaida macroura*, of North America. Many is the morning that I have awoken in Florida to its plaintive call of 'Oh, dear! Poor me!' It truly is a sad bird and so well named. The genus derives from a name created by Charles Lucien Bonaparte, a nephew of the French Emperor, Napoleon Bonaparte. In 1822 Charles had married his cousin Zenaïde, daughter of the former King of Spain, Joseph Bonaparte, and the young couple had gone with the exile to live in America, where Charles Lucien became an eminent ornithologist. When he named the dove, he did so for his wife, perhaps in the spirit in which Audubon annotated his portrait of the bird: 'I have tried, kind reader, to give you as a faithful a representation of two as gentle pairs of Turtles as ever cooed their loves in the green woods.' McCarthy also cites these lines in a fascinating chapter devoted to the symbolism of doves.

The species name *macroura* derives from the Greek *makros* (long) and *–ouros* (tailed), which reflects an earlier name, the Long-tailed Dove. Wikipedia reports that 20–70 million are shot each year, but that numbers stay high because the species breeds rapidly and many times a year. You can't wonder the poor little beggars are sad: they remember what happened to the Passenger Pigeon.

And, of course, the same Charles Lucien is the Bonaparte of the gull species.

Woods, stocks, rocks and rollers

Back in Britain we have three larger members of the Columbidae, which all belong to a third genus: they are the Woodpigeon, *Columba palumbus*, the Stock Dove, *C. oenas*, and the Rock Dove, *C. livia*. The last one is also known in its urbanised form as the Feral Pigeon. One thing which is immediately obvious in those names is that the words 'pigeon' and 'dove' have little real difference in usage. The truth behind this is a historical one. Lockwood traces the origins of 'dove' back to an old Germanic form which is based on the sound of the birds. He credits the medieval Church with bringing it into wider use after an earlier avoidance of it for reasons of Norse superstition. 'Pigeon' finds its way into English via Medieval French, but mainly with reference to the young birds, and is eventually more widely used after the seventeenth century.

Today there is a general feeling in the naming of the Columbidae that the larger species are pigeons and the smaller species are doves. The concept may well be supported by the fact that the word 'dove' is synonymous with peace and harmony. It is a wholly inexact distinction, as our own species show. What's in a name, after all?

At a local fair recently, somebody asked my advice on how to keep Woodpigeons off the garden feeders. It is a relatively recent phenomenon for this species to come into suburban and urban gardens in such numbers. In that it seems to have followed the example of the Collared Dove. Woodpigeons were certainly much shyer of humans when I was young: the number and use of shotguns and air rifles has fallen over the past decades. The Woodpigeon's name implies that it ought to be in a wood, though it is more a bird of marginal bushes and hedgerows, a habitat which gardens replicate quite well. In fact Gilbert White implied that it was far less common around Selborne in the eighteenth century than was the Stock Dove. Even today, Selborne is surrounded by areas of hanging woods (i.e. they grow on steep slopes). These woods contain many old trees, which provide the nesting holes required by the doves. The word 'stock' is synonymous with 'trunk' and relates to 'stockade', an enclosure built of tree-trunks – and by extension to the animals housed in the stockade. Their specialist nesting alone has been sufficient to ensure that their numbers have fallen considerably with the decline of woodlands and an increase in coniferous forestry. In 2000 it was estimated that Woodpigeon numbers in Britain stood at about six million individuals, by a long way the most numerous wild bird (though the species biomass is estimated to be only 75% of that of the heavily managed Pheasant population). However, winter flocks arriving from the Continent tend to sway the balance, making the Woodpigeon a serious crop pest in severe weather. Stock Doves were estimated at only 600,000.

YOUNG STOCK DOVES

Columba palumbus contains Latin names for dove and Woodpigeon respectively. *Columba oenas* uses the Greek *oinas* (pigeon). As for the convention of forming the common name of Woodpigeon as a single word, that form is used by the BOU and by a great many birders. The recommended International name is the Common Wood Pigeon.

The scientific name of the Rock Dove, *Columba livia*, is simply the Latin for Rock Dove. Historically the native habitat of Rock Doves was cliff faces, and these soon included the artificial cliff faces of human habitation. Over millennia, the species was semi-domesticated with pigeon lofts and dovecotes, and it underwent the sort of selective breeding which turned the Mallard into the farmyard duck. In 2011, a group of us stayed at a guest-house in Andalucia which was known as El Palomar. Behind it was the massive medieval dovecote which gave it the name. It once had the capacity to raise literally thousands of Rock Doves as a food commodity, and had for centuries supplied the Spanish military with pigeons for food and for communications, and guano for the manufacture of gunpowder. Today it is home to a few pairs of Lesser Kestrels and some Little Owls. On a smaller scale, a beautiful cylindrical dovecote still stands at Amesbury, Wiltshire, where it was constructed and exploited as part of a mixed farm. A number of such buildings are still found around the country. The remnant of this domestication is represented today by the Racing Pigeon enthusiasts on the one hand, and by the Fancy Pigeon enthusiasts on the other. There has long been a passion for ornamental birds of all colours and shapes: in Merrett's 1667 list he mentions a string of types, such as Culver, Jacobin, Runt, Helmet and Tumbler, and some of these names will still be found in the fifty or more types listed today, which include a Roller, a form which loops the loop for fun. This interest was exploited by Charles Darwin in his understanding of the processes of evolution.

Because the birds travelled in domestication to far-flung continents, I have seen them living comfortably near fast-food outlets in Florida, and in the airport at Sydney, NSW. In short, the modern Feral Pigeon is the descendant of domesticated birds which have reverted back into the wild and which still largely inhabit our towns and cities. Some of these have re-colonised the cliff faces of southern Britain, where the vast majority have reverted to the original pale-grey and dark-headed form.

In Florida and in Trinidad we encountered some of the tiniest members of the family, the sparrow-sized ground doves, while Australia's Diamond Doves were scarcely larger. The family seems to be much better represented in warmer climates, with Australia alone having two dozen of the world's 334 species, including three introduced birds. Of the natives, Banded Fruit Dove and Partridge Pigeon, *Geophaps smithii*, were my great prizes in Kakadu, the latter gaining its generic name from its ground-feeding habits and the vernacular name from its general resemblance to the partridges.

THE GREEN TIME BOMB: FERAL PARAKEETS

Several decades ago a number of escaped parakeets began to establish a successful breeding population on the southwest outskirts of London. The Rose-ringed Parakeet, *Psittacula krameri*, has since become an entrenched British resident species, slowly radiating out from that area. They are formally known in Britain as Ring-necked Parakeets. Clearly the difference between the two names we use is in the detail: the adult male develops a black neck-ring, which is often suffused with an edging of pink. *Psittacula* is merely a diminutive form of the Latin for parrot, which in turn derives from the Greek word *psittakos*. The specific name *krameri* was given by Scopoli to honour Wilhelm Kramer, a German-born naturalist who worked in Austria and died in 1765. Kramer was one of the first to adopt Linnaeus's binomial system and is also credited with creating the name Pratincole. To many people it seems a miracle that such an exotic species can thrive in the British climate, but the parakeets have a natural home-range in Asia and Africa which includes temperate areas at altitude, where they are naturally hardy. It also occurs as an escape in some other European countries: I recorded several of these in a park in Seville, Spain, and it is also one of many well-established feral parrot species living in Florida, where we saw it often.

ROSE-RINGED PARAKEET

The climate of Florida readily supports feral populations of a number of parrot species. Even more numerous in the Clearwater area was the Monk Parakeet, *Myiopsitta monachus*, of South American origin. A small population of this species was also established just to the north of London, but has probably now been eradicated. The generic *Myiopsitta* combines a corrupted form of the Latin *mus* (mouse) with *psitta* (parrot). The mouse element suggests that the parakeet is small and skulking, but it is rather that its relatively subdued coloration makes it less obvious than some of its kin. The hooded appearance of this grey-green parakeet with a whitish face and front leads to the names *Monk* and *monachus* (cf. Cinereous Vulture). This is another species which I have recorded in the wild in Spain.

Unfortunately both species are intrusive and have reputations in their native lands as serious crop pests. Rose-ringed Parakeet is a hole-nester and is already causing some concern for its aggressive, colonial use of available holes which would otherwise be used by native species, and it appears to be spreading inexorably into a wider area. Culling is now under consideration.

As a birdwatcher, I find them a colourful addition, but feel that it is only a matter of time before we see these intruders disturbing further the ecological balances in which some of our native species are already struggling.

The parrots are normally well represented outside Europe, with over 360 species worldwide and a further twenty-one of the related cockatoos: Australia and Trinidad have provided me with sightings of about one-tenth of the total number.

WELCOMED BY MEN, IF NOT BY BIRDS: CUCKOOS

It may come as a surprise to some to realise that the name Cuckoo is something of a latecomer to Britain, since there was an older, Nordic name, *Gowk* (and variants of that spelling) which was brought by the Vikings and which lingered in the north of England and Scotland until at least the seventeenth century. This form had much older Germanic roots: Lockwood postulates that the word was almost certainly onomatopoeic in origin, and condensed from two syllables to one by usage.

The word 'cuckoo' was a Norman French import as *cucu*, an evolution of the Latin name *cuculus*. It was first noted in 1240 in the round-song, 'Sumer is icumen in, Lhude sing cuccu!' which has long been associated with Reading Abbey (just five kilometres from where I am writing). Over the centuries, there were many variants of the spelling. Pennant, of course, was to settle on the current form, which then became definitive.

Even today, any sign of spring's arrival is greeted with pleasure. We welcome signs such as the first Swallow, or even the 'daffodil that comes before the swallow dares'. The robust and cheery song of the monastery welcomed a widespread harbinger of the end of the misery of winter. Nowadays, the Cuckoo is much less part of the scene and is probably unheard by most of the population, its significance as a symbol of spring waning with its numbers.

Today it is the Common Cuckoo, *Cuculus canorus*, the specific name meaning 'melodious'. It is part of a much wider family: in fact three vagrant species have been recorded in Britain. Great Spotted Cuckoo, *Clamator glandarius*, migrates in summer to the Mediterranean countries, but the occasional overshoot occurs. The generic name *Clamator* means 'shouter' and is supported by the *Collins Bird Guide* description as 'noisy and loud'. The specific name *glandarius* is like that of the Jay, though in this case the name has not so much to do with acorns as with the cork oaks which it frequents, since it is an insectivore.

Two American cuckoos have also been recorded here. Both belong to the genus *Coccyzus*, which derives from the Greek *kokkuzō*, meaning 'to cry like a Cuckoo' – except that neither of them utters a cry even a bit like the old world Cuckoo. The more frequent is Yellow-billed Cuckoo, *C. americanus*, while the scarcer one, Black-billed Cuckoo, *C. erythopthalmus*, has a name describing the red orbital ring of the eye (which is not red, should you see a juvenile).

The cuckoo family (Cuculidae) includes coucals and anis, which are not brood parasites. I have seen about ten of the family: the huge Channel-billed Cuckoo is mentioned elsewhere, but Australia also produced Pheasant Coucal and five other cuckoos (including three species in a single afternoon in the Botanic Gardens in Canberra – Pallid, Fan-tailed and Horsfield's Bronze).

FAN-TAILED CUCKOO

Trinidad's contribution was the grotesque Smooth-billed Ani, as well as a Striped Cuckoo (a rare example of brood parasitism in the Americas), and my favourite for its looks, the tiny, foxy-red Little Cuckoo.

None of those we heard produced anything like the familiar call of our Cuckoo: indeed several were variants on trills: the most distinct vocalisation came from the Common Koel, *Eudynamys orientalis,* which had just arrived in Darwin. The 'koo-well' call provides its vernacular name, while the generic speaks of 'strength': it is the size of a Carrion Crow.

Like ours, all the Australian species, the Coucal excepted, are brood parasites: I located one Fan-tailed Cuckoo by following the agitated behaviour of a pair of Olive-backed Orioles. On the Essex marshes I once counted the number of nesting Meadow Pipits, as a calling Cuckoo drew attack after attack by flying low over their territories.

STRIX AND TYTO: OWLS

The Canadian author Margaret Atwood is well known for her passion for conservation, and birds in particular, so I own freely that her novel 'Oryx and Crake' gave me the idea for that header.

If owls do not sit easily under the heading of raptors, they do at least answer to the collective term Strigiformes.

I was once told that Tawny Owls, *Strix aluco,* succeed in courtship because they have the wit to woo. That pun on their call and song makes nonsense of the generic name, which is derived from the Latin for 'screech owl'. In the simplest of terms, it is *not* the 'screech owl', to which I will return in a while. The specific name turns the Italian name *alucho/allocco* (Tawny Owl) back into a Latinate form. One of its folk-names is Brown Owl, but the name Grey Owl also existed. This arises from the fact that there are two colour forms of the species, which is apparently why Pennant created the compromise name Tawny Owl in 1768.

Screech Owl was a name used widely when I was young, and was locally often pronounced closer to the medieval form Scritch Owl. It distinguished the Barn Owl, *Tyto alba* from the Tawny's eerie, but pleasant call. The strangulated wheezing call of the Barn Owl is far from pretty and, when heard in the dark, can sound deeply sinister. When Barn Owls adopted church towers as nesting sites, they didn't endear themselves to humans by gliding, silent and white, through graveyards. For the unwary and superstitious, the unearthly wheeze was the last straw. Its alternative names of White Owl and Church Owl are more easily explained, but the term Barn Owl, which was first noted by Ray in 1678, has stuck. *Tyto* is from the Greek *tuto* (a night-owl) and *alba* is of course 'white'.

The screech owls of the Americas are *Megascops*. My one encounter with the Tropical Screech Owl, *Megascops choliba*, was in Trinidad, when our guide lured one close with a taped call. I have never seen such a malign scowl as on the face of that bird when the lamp told it that it had been fooled. (And there may be some on the faces of birders too at the thought of 'cheating' in that way. In my experience, however, I have seen the technique used only rarely and very sparingly.) The name *choliba* is an import from Spain, where it was the Aragonese name for the Scops Owl. (The latter, incidentally, is *Otus scops*, and that derives from the Greek *skopos*, 'a watcher'.)

The look on the face of that Trinidad owl could really have justified the superstitions of the terrified parishioner. We have to accept that owls have generally been profoundly misunderstood by the human race for those sorts of reasons. Even without the nocturnal shenanigans, the large forward-facing eyes and a profoundly unfathomable expression have always made us uneasy or overawed. We have a tendency to take the anthropomorphic approach to birds like owls, penguins and puffins, which have some human characteristics. There is no such thing as the 'Wise Old Owl', since they are actually not very intelligent, and, of course, they don't have malign intentions, unless they consider that you are too close to their nest. The photographer Eric Hosking lost an eye to one in those circumstances, but his monochrome pictures of Barn Owls became iconic and caught my youthful imagination.

But one Barn Owl did work a sort of magic on me, because my father often took me out with the gun and two dogs to shoot rabbits (myxomatosis hadn't been introduced, and we were still deep in post-war rationing, so this was subsistence hunting as well as pest control). One night he helped me onto a low oak branch and told me to sit still and quiet. I did, and was eventually joined, not three metres away, by a Barn Owl. I was just eight years old, and from then on was a birdwatcher. In *La Gloire de mon Père*, Marcel Pagnol tells the tale of an owl encounter, but this time in the mountains of Provence, when he was lost at a similar age. He encountered an Eagle Owl and was scared witless, but always retained a profound love of nature.

BARN OWL

Like those of the Eagle Owl, *Bubo bubo*, the 'ears' of Short-eared Owl, *Asio flammeus*, and Long-eared Owl, *Asio otus*, are no more ears than the 'horns' of the Great Horned Owl, *Bubo virginianus*, of North America: they are all feathers, of course, the real ears being hidden beneath the facial feathers and with no external evidence.

EAGLE OWL

A second fallacy is exposed by the Short-eared Owl, *Asio flammeus*, in particular, because not all owls are nocturnal hunters: some species hunt by day. The word *flammeus* in the name of that owl relates to the 'flame-colour' of the orange-buff patch on the wings. One intriguing archaic name for this species was the Woodcock Owl, which was explained by R. Bosworth-Smith in 1913 when he wrote: 'He is a bird of passage, which, appearing along with the woodcock in the autumn and disappearing with him in the spring, and flushed as he often is, like the woodcock, in boggy ground, and having the same kind of drifting zigzag flight, is known as the woodcock owl.' The two species are also linked in the legend that the tiny Goldcrest migrates by hitching a ride on one of those birds.

The last paragraphs introduce two new generic names, *Asio* and *Bubo*. The former simply means 'a type of eared owl', while the latter is the Latin for an Eagle Owl. While I have never seen the Eagle Owl in the wild, we did locate Great Horned Owl, *Bubo virginianus*, though not in Virginia: we found them regularly at Honeymoon Island in Florida, where they always occupied an abandoned Osprey nest. My only European sighting of a *Bubo* owl was Snowy Owl, *Bubo scandiacus*, but even that had not come from Scandinavia, as its name implies. It was in fact a known ship-assisted vagrant from the wilds of Quebec, which I saw sitting in Felixstowe Docks. And I didn't have to 'twitch' that bird, since I viewed it across the estuary during a visit to my old home town: a rarity on a plate. (I am not, by the way, interested in competitive listing, so what I see as a wild bird goes on my list, ship-assisted or not. It is *my* list. And if you want a really down-to-earth take on this topic, the comedian Alex Horne's *Birdwatching Watching* is a must read!)

Two of the larger northern owls belong to the same genus as the Tawny Owl. Great Grey Owl is *Strix nebulosa*, meaning 'misty, cloudy or dark'. The Ural Owl, *S. uralensis*, takes its name from the Ural Mountains, at the eastern edge of Europe.

The common name of the Northern Hawk Owl, *Surnia ulula*, is easily explained by its hawk-like flight and diurnal habits. *Surnia* is somewhat more controversial, since its origins are doubtful. It was given by Duméril in 1805, but various opinions either support it as a word of Greek origin or deny its validity. The specific name *ulula*, meaning 'screech owl', seems also to be the origin of the word 'owl' itself.

Europe's smallest owl is the Pygmy Owl, *Glaucidium passerinum*. The generic name derives from a diminutive of the Greek *glaux* (owl), while the specific name means 'sparrow-sized', though it is in fact a little larger than a House Sparrow.

Tengmalm's Owl (now known internationally as the Boreal Owl) is named after the eighteenth-century Swedish doctor and naturalist, Peter Tengmalm. Formerly classified as *Strix*, its current generic name *Aegolius* is taken from the Greek for 'screech owl'. The specific name *funereus* refers to the dark plumage, and maybe more so to the smoky dark brown of the young, but nonetheless the name sits well with the traditions of the churchyard.

(And finally, if Pennant could invent a name... The Torny Owl, I admit, was no accident – unless you consider that it fell into the brambles. Barn Oil, on the other hand was genuine typo. I know that W and I are a long way apart on the keyboard, so I can only assume that this one evolved subliminally as I typed. As always, I looked for a definition to fit the bird, and it came easily: the Barn Oil is a cure for squeaking rodents.)

OF FERN OWL AND GOATSUCKER: NIGHTJARS

Like the Cuckoo, the Nightjar was the sole representative of its genus in Britain, so the basic name evolved without adjectives. The root of the Willughby/Ray inclusion of the Nightjar in the category 'nocturnal birds of prey' was the fact that it had colloquial names such as Churn Owl, Fern Owl and Goat Owl. The bird's cryptic plumage gives it a superficial resemblance to the Tawny Owl, and, in fact, the two families are not that distant in relationship. However, their evolution has been greatly different, the nightjars becoming primarily insectivorous. Willughby and Ray seem to have ignored the fact that they did not have the 'crooked beaks and claws' of the owls.

The European Nightjar, *Caprimulgus europaeus*, began picking up curious folk-names in Ancient Greece, where the bird's habit of catching moths around the flock earned it the completely false reputation of taking milk from the goats. This myth lies behind the generic name *Caprimulgus*, which translates into the old English folk-name Goatsucker. The roots of the Latin form are *capra* (nanny goat) and *mulgere* (to milk).

NIGHTJAR

The modern name Nightjar, first noted in 1630 and formalised by Yarrell in 1843, reflects the fact that it 'jars' or 'churrs' in the night, the latter also giving rise to Churn Owl. That name, of course, brings a convenient coincidence between the sound and the large can (churn) in which farmers once collected their milk, goat or otherwise. Fern Owl came from the habitat plant: fern, or bracken.

Today it is the European Nightjar, because there is another in Spain, the Red-necked Nightjar, *Caprimulgus ruficollis*, which has no surprises in its name. There are also many other species throughout the world. We saw White-tailed Nightjar in Trinidad, which also provided us with sightings of Pauraque and Short-tailed Nighthawk, which are members of the same family, and of the less closely related Common Potoo and Oilbird. Together with the Frogmouth which we saw in Australia, all of these birds belong to the order Caprimulgiformes. I will return to some of these later.

THE DEVIL OF A DIN AND NOMADIC GREEKS: SWIFTS

Today the shrill screams of a party of Swifts probably brings more pleasure than fear, but the superstitious past failed to spare the crepuscular habits of a dark bird from acquiring a plethora of names on the theme of Devil Screech, Devil Shrieker, Devil Bird and the like. The 'Devil' was in the blackness, and that was also the sin of the Coot, to which the word was also apparently applied.

The name Swift, which is self-evident in meaning, was first noted by Charleton in 1668 and eventually standardised by Pennant in 1768. It is also swift in its visits to northern Europe, coming later and leaving earlier than many of our other summer migrants. With its close resemblance to the hirundines, in which there is another example of convergent evolution, it had also enjoyed such names as Black Martin, but swifts are in fact closer to the nightjars than to martins.

The species which breeds in Britain is, internationally, the Common Swift, *Apus apus*. The word *Apus* (footless) refers to the almost vestigial legs and feet, which are just large enough to permit the bird to cling to a vertical surface or to crawl into the nest. Since it spends almost its entire life in the air, and even mates on the wing, the legs are almost superfluous. Some years ago this was brought home to me when I found a Swift spreadeagled on my lawn: it was uninjured and had probably underestimated the amount of lift in the dead air of the enclosed garden. The tiny legs gave it no support on the complex horizontal surface of grass stems and its wings were stranded. A re-launch from the palm of my hand solved the problem.

Other species of the genus occur in Europe and some are vagrant to Britain. Of these the Pallid Swift, *Apus pallidus*, is paler, so the names are self-evident, while Little Swift's specific name, *A. affinis*, suggests that it has 'affinity to' the Common Swift. The Plain Swift, *A. unicolor*, could not be plainer. However, the White-rumped Swift, *A. caffer*, bears an archaic name which recalls its South African wintering grounds, but in a rather uncomfortable form. *Kaffirland* originally developed from an Arabic term for 'unbelievers', but history has distorted the term into a far greater racial insult. This seems a good case for waiving the 'oldest fool' rule on scientific nomenclature.

Alpine Swift, *Tachymarptis melba*, will often be found classified in the *Apus* genus. It is one of two species which are much larger than other swifts, but the resurrection of a 1922 genus has depended on more complex mitochondrial analysis. The 'new' genus consists of elements which add up to 'fast catching', which does seem a rather heavy idea. However, the specific name *melba* is a peach of an enigma. It seems likeliest that it is a compression of *melanoalba* or *melalba* (both meaning 'black and white'), since that is the way the bird appears compared to most other swifts, and since Linnaeus referred to those features in the note: *H. fusca, gula abdomineque albis* (a dark swallow with white throat and abdomen.) There are, after all, other examples of similar enigmas in names coined or confirmed by Linnaeus (and note that the swifts were then erroneously classified as swallows). From a British point of view, the Alpine Srift is a prized vagrant, and readily identified by its size and large white belly-patch. I saw my first one over the coast of Provence in 1973 and met them later in much greater numbers in Andalucia.

The American species, Chimney Swift, *Chaetura pelagica*, also appears on the

British list. In 2010, I was at a famous dairy chimney in Wolfville, Nova Scotia (arguably the World's smallest nature reserve), to watch the Chimney Swifts return to roost at dusk. How they manage the last-second vertical plunge is a wonder to behold. *Chaetura* is a name applied to a New World genus of swifts, others of which we saw in Trinidad. The word means 'hair-tailed'. The specific name *pelagica* is a bit of a surprise, however, because it has nothing to do with the bird crossing the ocean to get spotted in Europe. Rather, it originates from an Attic Greek word for swallows and swifts, deriving from the name of a nomadic tribe of Ancient Greece, the Pelasgi. Now that *is* obscure!

However, for those without good sea-legs who take pelagic birding trips, you will be glad to know that the word used in that sense derives from the Greek for 'a level sea', and its Latin version *pelagicus*.

LITTLE STUNNERS, OR THE HALCYON DAZE: KINGFISHERS

I referred earlier to the legend of Atthis, which gave rise to the specific name of the Common Kingfisher, *Alcedo atthis*. Lockwood suggests that the word 'kingfisher' itself may well have roots in the Fisher King legend from the Arthurian stories of Percival and the Holy Grail. The word started life as *fiscere* (fisher), but the form King's Fisher appeared as late as 1658, apparently with the inference of excellence at the skill of fishing. That spelling was to remain in use long after Pennant preferred the modern form in 1768, since it appears in Selby in 1833.

Once again the word for the one British species becomes the name of the worldwide family, and ours is now Common Kingfisher, *Alcedo atthis*. In the scientific form, the Latin name *Alcedo* is preferred over a previous usage of the Greek-based word *Halcyon*, which is now most familiar in the phrase 'Halcyon days'. Much of the vocabulary of the kingfisher family seems to stem from the legend of Alcyone and Ceyx, whose sacrilegious love-chat led to them being turned into kingfishers by the angry gods. *Alcedo* and *Halcyon* both derive from Alcyone's name, while one genus of kingfisher is named after Ceyx.

COMMON KINGFISHER

Nowadays the kingfishers are all assigned to the family Alcedinidae, but they were once divided into three families, loosely known as 'the river kingfishers' (Alcedinidae), 'the tree kingfishers' (Halcyonidae) and 'the water kingfishers' (Cerylidae). The European species belongs to the first group, while most of the Australian species, with two exceptions, are in the second, and the American species are in the third.

As the term 'tree kingfishers' suggests, this group is far less dependent on water, and some of the Australian species live in dry areas, with reptiles as their main food-source. These are discussed more fully in the chapter dealing with Australia.

The dubious line between 'river' and 'water' kingfishers in fact conveniently separates the nine New World species into the latter group (the former Cerylidae). The element *Ceryle* is a name linked to a water bird mentioned by Aristotle and others. My experience of these birds is mainly limited to Belted Kingfisher, *Megaceryle alcyon*, which was numerous in Florida in winter, but frequent enough in parts of eastern Canada in summer. The band is on the breast, while the name *Megaceryle* suggests that it is the largest of the group. By contrast the American Pygmy Kingfisher, *Chloroceryle aenea*, which we saw in Trinidad, is the smallest. The element *Chloro–* is from the Greek *khlōros* (green) and *aenea* means 'coppery-bronze', which two words neatly describe the back and underside coloration of the bird.

A STING IN THE TALE: BEE-EATERS

The European Bee-eater, *Merops apiaster*, is rare enough in Britain to have earned no folk-name, so the present vernacular name was coined by Charleton in 1668 as a translation of the Latin *apiaster*, which still features as the specific name. *Merops* means the same thing in Greek. My only encounters with these beautiful birds have been in Spain, where one of the highlights of a dawn outing was the arrival of a flock of about twenty chirruping shadows which came to settle on a nearby tree to await the first rays of the sun.

RAINBOW BEE-EATER

Blue-cheeked Bee-eater, *Merops persicus*, is on the British list, its specific name underlining the fact that it was a long way from Persia (modern Iran).

The Australian Rainbow Bee-eater, *Merops ornatus*, is no more beautiful than the European, but it has a more worthy name, though I feel that 'rainbow' works better than 'ornate'. We first encountered these at Fogg Dam in the Northern Territory, and subsequently in other locations in the area, but to me the most important one was in a patch of New South Wales woodland a fortnight later, the first spring arrival to demonstrate to me, from the northern hemisphere, a clear example of their 'upside-down' migration in a September spring, when birds which have wintered in the north travel south to breed. Meanwhile I had been exploring the various floral shrubs of the area and had seen evidence of the healthy insect life which was present to provide the food the birds would need.

PARADOXICAL BEAUTIES: HOOPOE AND ROLLER

The crowned tyrant...

In Ancient Greece, and later in Roman times, the Hoopoe was a symbol of tyranny, with its crest as a crown and a long, strong bill a symbol of dominance. It had Aristophanes and Ovid to thank for perpetuating its link to the legendary tyrant-king Tereus.

The classical roots of the word Hoopoe are reflected in the Latin and Greek forms of its scientific name, *Upupa epops*, which are both clearly echoic of its song: 'oop-oop-oop'. From those forms evolved first the French *Huppe* and later the English *Hoop*. The form Hooper first appeared in Merret in 1667 and was adopted as Hoopoe by Pennant in 1768.

The fact that the name travelled via French reminds me that the first one I ever saw was part of truly Gallic legend. One day in 1960, we were out exploring the chateaux of the Loire Valley: as our car passed in front of Rigny-Ussé (the chateau used by Perrault for the original telling of the Sleeping Beauty legend), a Hoopoe flew up from the lawns, a great butterfly of pink, white and black, offset by the turreted castle. It could not have been a more wonderful setting for such an exotic bird. On occasions they do overshoot during spring migration, so when one turned up quite close to home a few years ago, I had to admit that the lawns of a brewery make a somewhat less romantic setting.

Perhaps I would have been even more deflated if I had known then that one of the names listed by Charleton in 1668 was Dung Bird, after the bird's filthy nest. That is explained by the fact that the nestlings defend themselves against predators by a number of means, which include the secretion of foul-smelling oil and the ejection of the contents of their bowels.

... and the gaudy crow

The Roller has a much more southerly and easterly range in Europe than that of the Hoopoe, so its occurrences in Britain are much rarer. Like the Hoopoe, it is one of the most colourful and striking of European birds. It is bright and Jay-like, so there is no great surprise in the scientific name, *Coracias garrulus*, in which the generic part derives from a Greek word for a small crow, while the specific name recalls the (Eurasian) Jay, *Garrulus glandarius*, for its noisiness. The harsh display calls uttered by the birds are in fact rather reminiscent of the Jay. Modern taxonomy, however, places the species and its congeners much closer to the bee-eaters than to the corvids.

The common name derives from the bird's Lapwing-like display flight and, in that, links to one of the types of fancy pigeon. There are a number of other species of roller, all natives of Africa or Asia, and all birds of gaudy plumage.

GOOD REASONS TO WEAR A HAT: WOODPECKERS

Whenever we heard a woodpecker while out in the countryside, my father would warn me to keep my hat on.

The word 'woodpecker' is first seen in 1530, in the work of John Palsgrave (who tutored two of Henry VIII's children) and was eventually formalised by Pennant. As might be expected, a number of regional folk-names also existed, the most colourful of which seemed to be for the Green Woodpecker. Of these, the names Yaffingale, and variants of that idea, picked up the bird's call (still referred to as a 'yaffle'), but followed the structure of the word Nightingale: as Lockwood suggests, this has more than a touch of ironic humour in it.

The formal English names for the three British species have similar, but slightly different routes into the current forms. Greater Spotted and Lesser Spotted were first coined by Ray, and both shortened later by Pennant to 'Great' and 'Lest' (*sic*). While the first was accepted generally, the form 'Lesser' was reinstated by later authors.

LESSER SPOTTED WOODPECKER

This seems a good point to consider the inconsistent use of the adjectives 'great/greater' and 'little/lesser/least' in general. Earlier I mentioned the use of 'Greater' and 'Lesser' Flamingo: similar pairings are used with the two short-toed larks and the two yellowlegs, for example. 'Great' Crested Grebe is opposed by 'Least' and 'Little' Grebes, while the gulls are 'Great' Black-backed and 'Lesser' Black-backed. 'Least' Sandpiper has no one counterpoint name. Just more traps for the unwary. But let us get back to the main business...

Green Woodpecker was first used by Merrett in 1667, quoted by Ray among several other versions, and subsequently formalised by Pennant in 1768.

Inevitably legend creeps into the scientific naming, since *Picus viridis* (Green Woodpecker) gets its name from the Roman legend of Picus, King of Latium, who fell for a wood-nymph and was punished by Circe by being transformed into a woodpecker. Having recently seen Picus pursued by Nisus, the Sparrowhawk, I can vouch that this was not such a good outcome. Anyone who has witnessed such a scene will know that Picus has a distinctive and un-kingly panic-call.

While the OED states that the origin of the words 'peck 'and 'pick' lies in Latin, but of unknown origin, it seems probable that *picus* was in there somewhere, especially in the light of the French equivalent, the verb *piquer*. The word *viridis* simply means green.

Both Great and Lesser Spotted Woodpeckers are classified as *Dendrocopos*. The Greek *dendron* means 'tree' and *kopos* means 'striking'. It is not so difficult to work out which of the two birds is *major* and which *minor*.

Our fourth species of this family is the Wryneck, *Jynx torquilla*. The word Wryneck came from Merrett in 1667, was used by Ray, and then fixed by Pennant. *Jynx* derives from Greek *iunx* for the bird, but the origin of that word comes from a cry uttered during the grisly Ancient Greek custom of tying a Wryneck onto a string and whirling it around the head to bring back an errant lover. Little wonder that the poor bird has a wonky neck, as the English name implies. 'Wry' is a little-used word in modern English: it occurs in such as 'a wry grin', and means 'crooked'. The bird turns its head through a half-circle to make a threatening snake-like display, and it is that habit which lies behind both 'wry' and the specific form *torquilla*, which comes from the Medieval Latin *torquere* (to twist), and was created by the Venetian Theodorus Gaza in 1476.

WRYNECK

Europe offers several other species of woodpecker not seen in Britain, where the birding world is holding its breath for the appearance of the large Black Woodpecker, *Dryocopus martius*. *Dryocopus* is from the Greek *drus* (tree) and *kopos* (striking) – and a clever way of creating a word to repeat the concept of *Dendrocopos*, but with a variation for a different genus. Behind the word *martius*, a type of woodpecker, is the Latin word *martulus* (hammer) (cf. modern French *marteau*). The form *Martius* was also used adjectivally for Mars, the Roman god of war, and has links to the Frankish ruler Charles Martel (Charles the Hammer), so called for his exploits in crushing the Moorish invasion at Poitiers in the year 732 CE.

In *Picus canus*, the Grey-headed Woodpecker is treated to the same dull grey specific name as the Common Gull. In the names of the Syrian Woodpecker, *Dendrocopos syriacus*, and Middle Spotted Woodpecker, *D. medius*, there are no mysteries, though the former's range now extends well into a dozen countries of southeast Europe. In the case of the White-backed Woodpecker, *D. leucotos*, the name appears to be a compressed version of *leuconotos* (white-backed), to which Bechstein amended the name three years after his original 1802 version. Presumably historical precedence prevailed over the amendment.

Three-toed Woodpecker, *Picoides tridactylus*, is something of the odd man out, as the name *Picoides* (resembling *Picus*) suggests. Most other woodpeckers have four toes, so the absence of one toe gives rise to 'three-toed' and *tridactyla*. This one is now the Eurasian Three-toed Woodpecker, following a split from the American form, now *P. dorsalis*, which name draws attention to the white back.

On this occasion, I cannot refer to the Australian woodpeckers, since they have none: the cockatoos and parrots fill that ecological niche. The Americas, however, have plenty, and I have so far seen five in North America and four in Trinidad.

Of these the most frequent has been the Northern Flicker, *Colaptes auratus*, which has a fascinating name born from the golden shafts of the primaries and the golden under-tail of the bird, which flicker like flames as the bird forages in the sun. In Florida I watched one foraging in moss in unused holes on a concrete power pole: it seemed a dangerous and foolish place to exercise the *Colaptes* element of its name, which means 'chiseller'. There is a red-shafted form of this species to the west of the Rockies. It is a distinctive woodpecker with a buff-brown mantle and white rump, which together make it look like a drunken (Eurasian) Jay as it flies off in typical woodpecker undulation. To make matters more curious, its call is like a high-pitched Green Woodpecker. Stranger still, as the State Bird of Alabama, it is popularly known as the Yellowhammer. In this case the 'hammer' element has to be taken literally, but in the name of the British bunting that element is a corruption of an old Germanic word meaning just 'bunting' (see below). The species also forms part of a record sighting which I had in the dawn light of Nova Scotia: on the branches of a single dead pine were two Northern Flickers and a family of four Downy Woodpeckers – all in the scope at once!

Another of my favourites is, of course, the magnificent Pileated Woodpecker, *Dryocopus pileatus*. It is huge by British standards, though well camouflaged in dense Everglades mangroves, as I discovered to my chagrin the first time I heard one. When I saw another further north, I was blown away. This is the Woody Woodpecker of the cartoons, which wears a distinctive scarlet cap – the *pileus* of its name. The story of its relative the Ivory-billed Woodpecker has become a legend in its own right. If ever a bird could be willed back into the sight of humankind this is the one for the Americans. Sighting was claimed in Arkansas in 2004; since then, expeditions to re-find it have taken on the flavour of Indiana Jones, and huge debates have raged, but it still just may still be there. The very name is causing a nation's ornithologists to hold their breath.

We saw its showy cousin, the Lineated Woodpecker, *Dryocopus lineatus*, in Trinidad, sporting the white 'go-faster' stripe of its name on its black bodywork, but my favourite was the Chestnut Woodpecker, *Celeus elegans*, simply because we do not have any woodpeckers the colour of a Red Squirrel. Curiously, *Celeus* means 'green woodpecker', yet, as far as I can see, none of the birds of this genus is even vaguely green!

THE BISY LARK, MESSAGER OF DAY

The spelling is Chaucer's, not mine. It occurs in 'The Knight's Tale' and has stayed with me since I studied that text many moons ago. It tells of the glory of sunrise and is one of my favourite images in bird poetry. Shelley, of course, awarded the bird a whole ode, and Vaughan Williams wrote a beautiful tone poem, 'The Lark Ascending'. In short, the role of the Skylark has always been a symbolic one, lifting the eyes, ears and spirits of the observer towards the heavens, as it rises on shimmering wings and trilling song to become a minute speck before swirling back down to its territory. I recall one particular sunny childhood morning when my sister and I sat quietly listening to that wonder for what seemed hours. She too has always loved her birds, and we both worry for those who live divorced from nature.

Lockwood points out that the word 'lark' has very old roots in the Germanic languages. It was Ray who created the term Skie-Lark in 1678, but he based the idea on the word *Himmellerch*, used by Gesner in 1555. Pennant found the idea useful as a specific name in its current spelling. Curiously, the word Woodlark is much older, being first used in the fourteenth century, and eventually occurring in the current spelling in Merrett in 1667, after which it was the standard name.

The scientific name of the Skylark is *Alauda arvensis*. Since *Alauda* is the Latin for lark there would seem to be no great surprise in the word, but Pliny apparently said that the word was Celtic and meant 'great songstress'. The specific name *arvensis* means 'of the fields'. Woodlark is *Lullula arborea*, the generic name being contrived from *Lulu*, a name created by Buffon in the late eighteenth century to describe the bird's call. The specific name means 'arboreal, tree-dwelling'.

The one other regular species in Britain, though as a winter visitor, is the Shore Lark, *Eremophila alpestris*. The name is suitable because most of those we see are on beaches and dunes. However, the generic name tells us that these are 'desert-loving' birds, while the specific name, *alpestris*, tells us that it is also a species of high altitudes. With all that confusing geographical data to hand, it has to be said that the least reliable element is the word 'shore', since that applies to only part of the population, while others may never see a shore. It has been argued that the American vernacular name, Horned Lark, is preferable, in that it describes a feature of the bird's appearance (though the 'horns' are of course feathers). The flaw there is that only the breeding male wears this feature, but Horned Lark is now the preferred international name.

Further south in Europe, other lark species occur: Andalucia produced a healthy crop of Crested, Thekla, Short-toed, Lesser Short-toed and Calandra larks. Of those, the Short-toed Lark is, quite logically, better contrasted to Lesser by use of the international name Greater Short-toed Lark, and I see no reason why the BOU should maintain the shorter name, especially as the two can be found together in many areas. Thekla Lark has a mysterious-looking name, but Thekla was the name of the daughter of the nineteenth-century naturalist, Christian Brehm. Calandra Lark takes its name from the Greek *kalandros* for the species.

As for their scientific names, Crested and Thekla are both of the genus *Galerida*, meaning crested (from the Latin *galerum*, 'bonnet, cap'). The specific name of the former, *cristata*, simply means 'crested', while *theklae* is self-evident.

THEKLA LARK

The generic name of the Calandra Lark, *Melanocorypha calandra*, means 'black lark', which is explained by the fact that it shares the genus with the Black Lark, *M. yeltoniensis*, which has appeared in Britain on the odd occasion. On the other hand, Calandra Lark can justify that name in part, because the blackish under-wings of the bird in flight can look intensely black in strong light. On the ground, however, it is a sandy brown and white bird with a black 'collar'. The second element of that word, *–corypha*, is based on another Greek word for lark (or an unidentified bird). The word *calandra* appears also in the name of the Corn Bunting, and is linked to the larks by an ancient and ongoing Mediterranean habit of catching larks and other small birds for food.

The Eastern equivalent of Calandra Lark is Bimaculated Lark, *Melanocorypha bimaculata*, which has been recorded only rarely in Europe. 'Bimaculated' means 'double-spotted', which is presumably a reference to the two black patches at the sides of the throat, a variable feature which it shares with Calandra Lark.

Greater Short-toed Lark is *Calandrella brachydactyla*, the generic form being simply a diminutive form of *calandra*. None too surprisingly, the specific name means 'short-toed'. Lesser Short-toed Lark is usually classified in the same genus, though the BOU and IOC lists now have it as *Alaudala rufescens*. That genus is a diminutive of *Alauda*, so the more correct form would be *Alaudula*. Perhaps more importantly it is *A. rufescens* (reddish), which description is only applicable to the nominate form found in Tenerife. The fawn-brown Spanish subspecies is *A. r. apetzii* (after the nineteenth-century German naturalist J. Apetz).

Dupont's Lark, *Chersophilus duponti*, is the one European breeding species which has escaped me – and a great many others, I believe, since it is reputedly hard to find, even in the semi-desert country which its generic name *Chersophilus* implies. The word is from the Greek *khersos* (barren land) and *philos* (lover). It was named by Vieillot after Léonard Dupont (1785–1828), a French naturalist and collector.

(This bird may well fit my very genuine typo, the Shylark. The bird is rarely seen, of course, and consequently has never appeared on any menus.)

THE SWALLOW DARES: HIRUNDINES

This is an out-of-context phrase from Shakespeare's A Winter's Tale.

As might be expected with such closely related birds, the words 'swallow' and 'martin' represent the dual roots of modern English. 'Swallow' is of Germanic origin, while 'martin' is a later introduction from French.

Lockwood traces the roots of swallow to a word meaning 'cleft stick', a reference to the forked tail. It was used in Old English as *swealwe* and later as *swalowe*. Before the advent of martin, local names prefixed the word swallow, creating forms such as Bank Swallow.

In recent years there has been a wider use of the international name, Barn Swallow, which now distinguishes the familiar bird, *Hirundo rustica*, from the Red-rumped Swallow of southern Europe. The association of the words 'barn' and 'swallow' is by no means an Americanism: the species has long nested in farm buildings in this country. In 1774 (Letter XVIII to Barrington) Gilbert White wrote: 'the swallow, though called the chimney swallow, by no means builds altogether in chimnies, but often within barns and out-houses .' He then goes on to recount the details of how the birds do in fact nest in chimneys and marvels at the dexterity they show to earn that name. Many years ago, my mother could quite readily have called them 'farmhouse swallows', as she discovered when spring-cleaning her home. One fine day all the windows were open in the bedrooms, and in no time the returning Swallows showed that they had a taste for comfortable quarters. It took her some time to scurry around shutting windows, and longer still to wash off the mud foundations which the birds had made on the picture-rails in every single bedroom.

The specific name is even more necessary in North America, where there are some six other species which bear the name 'swallow'. There are also several subspecies of Barn Swallow worldwide, the American being *erythrogaster*, meaning 'red-bellied', since this is a much more strongly coloured race.

Red-rumped Swallow is the only other bearer of the name in Europe, but it rarely strays to Britain, being more of a Mediterranean species. Its scientific name, *Cecropis daurica*, is intriguingly different. Firstly the generic, created by Boie in 1826, relates to an Athenian tribe whose name was derived from that of

Cecrops, the city's founder, and reputedly a mythical half-dragon or half-fish. The bird was first described and named by Erik Laxmann, a Finnish-Swede, who in 1769 obtained a specimen during an expedition in Russia. This accounts for the rather improbable use of the specific *daurica*, which refers to the Russian region of Dauria, to the east of Lake Baikal in Siberia. That name also appears in such as Daurian Jackdaw. The swallow's distribution is throughout Asia, though it seems possible that the specimen collected by Laxmann was well north of its normal range. The European form is in fact a subspecies, *C. d. rufula*, meaning 'rufous'.

'Martin' entered the English language much later than 'swallow'. It first appears in English in 1450 and develops variants such as *martlet*, *martnet* and *martinet*, all based on pre-existing French diminutive forms. For some centuries Martin was equally used to designate the Swift, which was sometimes the Black Martin. After Ray in 1678, the word Martin, used alone, designated the House Martin, *Delichon urbicum*, and the word Swift was used in the modern sense. The 'House' element is first seen in Gilbert White, in 1767 (Letter X to Pennant). His monograph on the species was one of the first in-depth species studies (Letter XVI to Barrington). This form echoes *urbicum* (town-dwelling), but was not taken up formally by the ornithologists until the BOU's 1923 list: it appears in Coward in 1926 as House-Martin, but in BOU's 1883 list, and in Hartert in 1912, the single word Martin had been the norm. Curiously though, the term Sand Martin had been in general use since Charleton had noted Sand, or Bank Marten (*sic*) in 1668, and the current form had been formalised by Pennant in 1768. The old English name, Bank Swallow, travelled to America, where it now provides one of those traps for the British birder.

I discussed earlier some aspects of the use of *Hirundo*/*Chelidon* for the Swallow, and *Delichon* for the House Martin. There are several other generic words within the family, three of which occur below. All members of the family are sometimes referred to as hirundines.

HOUSE MARTIN

The Sand Martin, *Riparia riparia*, has a name that relates closely to its habitat: *Riparia* is from the Latin for river-bank, and, of course, its vernacular name suggests a preference for sandy banks in which to dig its nest-holes. In like vein the specific elements (Crag and *rupestris*) of the name of the Crag Martin, *Ptyonoprogne rupestris*, tell us that it is 'rock-dwelling'. I had little doubt of that as they swirled round sheer rock-faces in Andalucia. The generic name, *Ptyonoprogne*, is descriptive, since the first element derives from the Greek *ptuon* meaning 'fan', which is a reference to the distinct stubby triangular tail. The element *progne* is from Procne, sister of Philomela, and one of those many mortals to suffer transmutations at the hands of the Greek gods, in this case into a Swallow (see the notes on Lapwing).

Procne appears again in the genus of the Purple Martin, *Progne subis*, of North America. The specific name *subis* seems to relate to a legendary bird of great courage: indeed the martin does appear to have a reputation for chasing off predators. It is unclear whether its long association with humans was encouraged for that reason or for the purpose of controlling harmful insects around settlements. Native Americans hung out hollowed-out gourds around their villages for the birds to nest in and the practice was later copied by settlers. Today, provision of nesting complexes has developed into a major hobby industry, and the martin would probably not survive without such help. Its site-loyalty made it the ideal subject to be the first species tagged by geolocators during studies in Canada.

The wordiest name among the America swallows is that of the Northern Rough-winged Swallow, *Stelgidopteryx serripennis*. What a mouthful! A few used to pass through Florida during our spring visits, and we saw the southern equivalent in Trinidad. This is a name which really overdoes the description of one feature: *Stelgidopteryx* tells us in Greek that the wings are like 'scrapers', while *serripennis* is the Latin for 'saw-winged'.

For me the most effective name belongs to the tiny Fairy Martin, *Petrochelidon ariel*, of Australia. It is not the generic name (rock-swallow) which attracts me, but rather the specific name . It refers to Ariel, a spirit of the air, who makes an appearance in Shakespeare's *The Tempest*. I think no scientific name could fit a bird better.

A BIT OF A LARK? PIPITS

It is quite unsurprising that the word 'pipit' has its origins in the call the bird makes. Neither is it a surprise to learn that pipits were once known as a species of lark, since they do resemble small larks. Forms such as Titlark and Meadow Lark preceded Montagu's use of Pipit Lark in 1802. Bechstein had already separated pipits from larks in 1795. In 1833, Selby established the three separate names Meadow Pipit, Rock Pipit and Tree Pipit. There were precedents for the first one in the word *pratensis*, used by Merrett in 1667, and in such usage as Meadow Lark. For the second there was Montagu's Rock Lark, which was also a traditional name. The third, Selby created from an observation by Montagu that the bird 'rarely alights on the ground without previously perching on a tree'. Yarrell standardised these forms in 1843. However, the status of Rock Pipit was reviewed by the BOU in 1987, when the former race *spinoletta* was split to become the Water Pipit. The North American Buff-bellied Pipit, *Anthus rubescens*, a rare vagrant to Britain, was also separated from Rock Pipit at that time.

BUFF-BELLIED PIPIT

The pipit genus, *Anthus*, was taken from a small grassland bird mentioned by Pliny, while *anthos* was used by Aristotle for what might have been a Yellow Wagtail. Almost inevitably there is a Greek legend and a transmutation lurking behind the word: Anthus was killed by his father's horses and became a bird which neighed like a horse and fled from them. Having narrowly missed being flattened by a runaway greengrocer's horse as a child, I can sympathise with him.

The specific names *pratensis* (Meadow) and *petrosus* (Rock) are simple translations, but *trivialis* works differently for Tree Pipit, since it means 'common' and seems rather an odd choice for a migrant species which is absent for much of the year. The name for the Water Pipit, *Anthus spinoletta*, derives from a delightful Florentine name, *spipoletta* (pipit).

I have seen just one pipit species outside Europe, Australia's only pipit. That species changed its name between my two visits, because it was known in my 2000 field guide as Richard's Pipit, *Anthus novaeseelandiae*. The situation has constantly been reviewed since that date: Richard's Pipit is now *A. richardii*, and for a time the Australian form remained a race of *A. novaeseelandiae*, but known as the Australian Pipit. The dual nationality has now been resolved by further splits of the New Zealand and Australian forms. What I first listed in 2000 has now changed both elements of its name to become Australian Pipit, *A. australis*. What is more, I recently added a vagrant of the Siberian form of *A. richardii* to my British list, which means that I have now seen Richard's Pipit for the first time – twice! The editing of a life-list tends to be a matter of continual maintenance. As for Richard, he was a French citizen of Luneville, commemorated by Vieillot in 1818, so the Anglicised pronunciation of the name is technically incorrect.

Other vagrant pipits in Britain include the widespread Tawny Pipit, *Anthus campestris*, whose name means 'of the fields'; Red-throated Pipit, *A. cervinus*, whose name suggests it is 'coloured like a stag' (an invention of Pallas in 1811); and Olive-backed Pipit, *A. hodgsoni*, which is named after a nineteenth-century

English collector and diplomat, Brian Hodgson. The nineteenth-century British zoologist Edward Blyth lends his name to Blyth's Pipit, *A. godlewskii*, which also records W. W. Godlewski, a nineteenth-century Polish naturalist who was exiled to Siberia. Such dual-naming seems rare, but it echoes the case of Latham's Snipe, *Gallinago hardwickii* (cited earlier), and Radde's Warbler, *Phylloscopus schwarzi*, which occurs later in this book, among the warblers. The extreme northerly range of Pechora Pipit, *Anthus gustavi*, accounts for its vernacular name, which is that of an area of sea in the Russian Arctic, while a nineteenth-century Dutch ornithologist, Gustaaf Schlegel, is commemorated in the scientific name and, unusually, by the use of his forename.

A TWIST IN THE TAIL: WAGTAILS

Before 1510, the wagtails were known as 'wag starts', the second element being an old Germanic word for tail, still used in the name of the redstarts. Wagtail became the normal usage, with the now-familiar adjectival additions of Pied, Yellow and Grey for the three British species.

The generic name, *Motacilla*, Jobling explains, is a little confusing. Strictly it derives from *muttēx*, a Greek word for a small bird mentioned by Hesychius, to become the Latin for wagtail, as used by Varro. However, in the Middle Ages, there was a misinterpretation of the word as a form of *motare* (to move) plus *–cilla*. The latter element was assumed, erroneously, to mean 'tail' and that error was perpetuated – a wag with a false tail, so to speak.

If ever a bird was hard done by in its name, it has to be the Grey Wagtail, *Motacilla cinerea*. This is arguably the most elegant of the three common British wagtails, and its wash of lemon yellow is so beautiful that the emphasis on its greyness seems unforgiveable. The fact that *cinerea* underlines the greyness merely rubs salt into the wound. In French the emphasis is at least on the bird's habits, since it is *Bergeronette des ruisseaux* (the Stream Wagtail.)

GREY WAGTAIL

In fact, the French are much more sensible with their use of grey, reserving that epithet for the White Wagtail, *Motacilla alba*, which is inaccurately named in both English and Latin. The 'whiteness' merely reflects the fact that it is a paler bird than the more striking patterns of the British subspecies, the Pied Wagtail, *M. alba yarellii*. The trinomial there is from William Yarrell, the nineteenth-century naturalist.

Since I have introduced the French word *bergeronette*, I would suggest that it was probably first applied to Yellow Wagtail, *Motacilla flava*, since, as most birdwatchers know, the best place to find the species on passage is around the heels of livestock. The word is a folksy extension of one used by Marie Antoinette when she and her ladies in waiting played at shepherdesses (*bergères*) during the heady pre-Revolutionary days of Versailles. In short, the wagtails were seen as 'tiny shepherdesses'. However, I have never seen a breeding male in a spring meadow without it conjuring the thought that buttercups are mobile. There is clearly something about them which provokes metaphor.

The British Yellow Wagtail is *Motacilla flava flavissima*, literally 'the most yellow of the yellow wagtails' (though *M. f. lutea*, with a 'saffron yellow' head, might rival it). There are a number of other races, most of which have darker heads: we saw the handsome blue-grey headed form, *iberiae*, in Iberia, or to be more precise, in the Coto Doñana, where there were a great many in April. This is one case where the *Collins Bird Guide* cheerfully lists and illustrates a dozen subspecies. Given the importance of John Ray in this narrative, it is good to find that the Yellow Wagtail was often known as Ray's Wagtail during the nineteenth century. The name was used by both Yarrell and Morris.

The vagrant Citrine Wagtail, *Motacilla citreola*, is 'lemon yellow' by name, though the yellow is restricted to the head and undersides.

I have never recorded a wagtail species outside Europe, unless you count Willie Wagtail, which happens to be a species of Fantail and was the first native Australian species I ever identified. More of that later...

BOBBING UP AND DOWN LIKE THIS: DIPPER

When I was a boy my home football team was nicknamed the 'Shrimpers'. Its warm-up song was 'Sons of the Sea', the refrain of which has given me the title above.

The word Dipper appeared as early as 1388 and may have been applied more for the bird's bobbing movements than for its habit of feeding under water. In fact the scientific name *Cinclus* may support that idea somewhat, since it derives from the Greek *kinklos*, which was used by Aristotle for a small tail-wagging bird. However, the name Dipper was rivalled for some centuries by the term Water Ouzel, which was indeed preferred by both Ray and Pennant. The word 'ouzel' is traced by Lockwood back to very ancient origins and was the original name of the Blackbird, *Turdus merula*. Dipper was used by Selby and Fleming in the early nineteenth century and became standard after Yarrell in 1843, though his version, Common Dipper, was eventually shortened. However, Water Ouzel lingered in some formal usage until well into my lifetime, since I recall it well in some of my youthful reading: it certainly occurs in Henry Williamson's 1927 novel *Tarka the Otter*, and, it seems, is still widely in use as an alternative name for the American Dipper, *C. mexicanus*.

Allowing for the fact that Williamson may have been seeking a certain element of romantic colour in his use of the archaic form, it is interesting that Coward's 1926 list is at the other end of the spectrum, with three forms under the names Black-bellied Dipper, British Dipper and Irish Dipper. The first of these is, of course, the nominate race, which occurs in Britain as an occasional vagrant from mainland Europe. The second is the British subspecies *C. c. gularis*, and the third is the Irish subspecies, *C. c. hibernicus*. The trinomial *gularis* is of little value, since all races have a distinctive 'white throat'. Since Hibernia was the Roman name for Ireland, *hibernicus* is more helpful.

BLACK-BELLIED DIPPER

This seems a good point to look a little closer at what the BOU/Coward list actually represents. Clearly, after much late-nineteenth-century disagreement on the topic, the BOU had espoused the trinomial system with relish, as this detailed entry shows. Coward's work, *The Birds of the British Isles*, was a three-volume reference, rather than a field guide, and perhaps should be compared with a work such as the monumental *Birds of the Western Palearctic*. There is a less overt use of subspecies in most modern handbooks: in the *Collins Bird Guide* the illustrations often tell us more than the text about the matter, but only where there is a clear visual difference. Yet it is a matter of great interest to many 'birders' and 'listers', especially those who wish to take insurance against future 'splits'. The use of such jargon words reminds me that Coward uses the word 'wanderer' for the Black-bellied form, where 'vagrant' would be today's usage.

Today the species is known internationally as the White-throated Dipper in order to distinguish it from four other species. Two intriguing questions arise with these birds. The first is why they bob, and the answer is still more in the realm of theory than not. Adjustment of the head helps to triangulate the potential prey, but that does not necessitate moving the whole body, as the herons show. One theory is that it indicates liveliness and the ability to react quickly if threatened, while another suggests that the movement blends with rippling water. The second mystery is more easily explained: they can walk under water because their bones are solid material, rather than the lightweight hollow-strutted bone of a normal flying bird.

BOHEMIANS AND CEDARS: WAXWINGS

There are few garden visitors more attractive than the waxwings. In early 2013 I at last had five in my own garden, but I had previously been lucky on three occasions to see some in gardens in Florida, Quebec and Nova Scotia. Mine were Bohemian Waxwings, *Bombycilla garrulus*, but the others were the smaller Cedar Waxwings, *B. cedrorum*. In Florida they were part of a huge wintering flock, but those in Quebec were stragglers still moving to breeding grounds in June after an exceptionally late spring. The third sighting was of one singing in breeding territory in July.

The name *Bombycilla* was originally created for the Bohemian Waxwing, which occurs both in Europe and North America. *Bombyx* is the scientific name of the silk-moth, and *–cilla* is 'tail'. The compound word appears to be a translation of the German name *Seidenschwanz*, meaning 'silk-tail'. That name is mentioned in Linnaeus's entry in the *Systema*, where there is evidence of an English name of that form, undoubtedly a mere translation of *Bombycilla*. The current Swedish name, *Sidensvans*, echoes the German form.

The word *Bombycilla* seems to have been a creation of Brisson in 1760 and was adopted later by Vieillot, because Linnaeus himself had (erroneously) classified the species as *Garrulus*. That name is explained by a BOU note of 1915: 'applied by Linnaeus to the Waxwing because of its fancied likeness to a Jay' (which is *Garrulus glandarius*). The word actually means 'chattering', which hardly describes the bell-like communication calls of the rather quiet Waxwings. Linnaeus's entry includes some variants of that idea, such as Buffon's *jaseur* and Latham's rather attractive Waxen Chatterer, but such names were ultimately meaningless.

While the 'silk-tail' element is based on one feature of the gorgeous European bird, the English name, Waxwing, is a reference to another of its special features, the waxy red blobs on the wing.

The vernacular adjective, Bohemian, is not much used in Europe, where there is only one species. It conveys the sense of 'Gypsy' or 'Romany', to describe the bird's wandering habits. As discussed earlier in relation to the Turkey, such names were loosely applied and often with little sense of geographical accuracy. Indeed, the very term Gypsy derives from the word Egyptian.

In Britain the wandering habits of the Bohemian Waxwings are unpredictable: we get 'waxwing years' because their arrival is sporadic, quite unlike the reliability of other Scandinavian species such as the winter thrushes. The term used for the Waxwing movement is irruption rather than migration, since the sporadic movement depends on a combination of breeding success and food sources, which do not always promote movement in our direction. In the winters of 2010/11 and 2012/13 the blighters were everywhere – and at last in *my* garden!

THE ACCENT IS ON TRADITION: DUNNOCK

The familiar word Dunnock uses a suffix deriving from Old English to describe 'a small dun-coloured bird'. Other parallels – Tinnock, Ruddock and Wrannock – have become obsolete, but this remnant name is now guarded jealously against the pressure from an alternative 'book-name'.

The word is first recorded in the late fifteenth century as *donek* and eventually in 1824 in the current spelling. This homely name existed in parallel to the term Hedge Sparrow, which first appeared 1530. The latter name was used by Merrett, Ray and Pennant. In 1825 Selby introduced the term Hedge Accentor, the latter word being an 1803 invention of Bechstein. It is based on the Latin *cantor* (singer), and was intended to create a family name equivalent to that of 'warbler'. That name was used by Yarrell, but Hedge Sparrow was preferred by the BOU in 1883 and by Coward in 1926, and it was widely used until well after World War 2. The problem with the name was a fairly obvious one, since, in Britain at least, 'sparrow' had become increasingly narrow in its application and was no longer a catch-all word for small brown birds. The name was changed formally to Dunnock in 1949 at the suggestion of Max Nicholson.

DUNNOCK

I suspect the story is not over yet: although the IOC uses Dunnock, Birdlife International still lists it as Hedge Accentor. It seems only a matter of time before that name is back on the agenda. With only two species of accentor in western Europe, and the Alpine Accentor a very rare vagrant to Britain, the idea has never found great favour here, especially since the remaining eleven species of that genus are found much further east.

In scientific terms, Dunnock is *Prunella modularis*. The generic name was created by Gesner in 1555, based on the German *Braunelle*, meaning 'little brown bird', and therefore has exactly the same sense as 'Dunnock'. The word *modularis* means 'melodious' (from the Latin *modulus*, melody), a tribute to the bird's quiet, sweet song. The Alpine Accentor is *Prunella collaris*, so named for a delicate white 'necklace' which shows up in fresh plumage.

One other interesting aspect of the name *Prunella* is that it coincides with that of a plant genus of the mint family. Such duplication is permitted under the rules of the Linnaean system, provided that the organisms named are not within the same Kingdom. In any case, no binomen may be exactly the same as another, so there is no possibility of *Prunella modularis* being confused with the plant *Prunella vulgaris* (Self-heal). A similar link between the Ruby-crowned Kinglet and the marigold occurs later.

ALIAS JENNY AND ROBIN (WRENS AND ROBINS)

The curious thing about the naming of birds is that so much was originally done in Europe. It is therefore ironic that the first-named species of the wren family, the Wren, *Troglodytes troglodytes,* should be the only wren to be found outside the Americas. A few years ago, birders in Britain were a bit put out to find that someone had stuck the word 'Winter' in front of the name of a familiar and unique bird. For the older generation, who remembered the jaunty image of the Wren adorning the old farthing (a pre-decimal coin in use until 1960), this bird was a national icon. It turned out that it was a 'blame the Yanks' moment, since the bird had been named within the international family and was considered to be the same species as an American one and therefore took its name. The simple explanation, which reconciled me to the idea, was found in Bruce Campbell's *Dictionary of Birds,* where he pointed out that the occurrence of many wren species in North America indicated that region as the original home of the family, and that one species had invaded Eurasia. Oops! All those ruffled feathers about 'our' bird, the British Jenny Wren. 'Unique' suddenly became 'odd-one-out'.

EURASIAN WREN

The invasion occurred a very long time ago, of course, and probably more than 4.3 million years ago via the Aleutian Bridge, an ancient land-link between northwest America and northeast Asia, and then westwards. Very recently, however, the matter has been reviewed yet again. American Winter Wren has been split into *Troglodytes hiemalis* and *T. pacificus* (Pacific Wren), leaving the Eurasian species to be called, not too surprisingly, the Eurasian Wren, *T. troglodytes,* which seems to make good sense.

However, there are twenty-eight recognised subspecies within the Eurasian range, and one of the unresolved matters of avian science is whether any of those should be split into full species. It seems very unlikely that the situation will remain static. Interestingly, the British and Irish mainland race is *T. t. indigenus,* while there are four more recognised subspecies on outlying islands, of which St Kilda Wren is the best known. The nominate form is that of continental western Europe.

The Americans have a long list of wren species, including House Wren, Cactus Wren, Rock Wren and Marsh Wren, the very names underlining the adaptations of this family. Surprisingly, I have found it quite hard to find a Winter Wren in North America, and have seen just one in Nova Scotia. Carolina Wren was a lot easier in Florida, where I saw it several times. The Southern House Wren showed up for me in Trinidad. There are so many of the family that a new species, the Antioquia Wren, *Thryophilus sernai,* was identified in Colombia as recently as 2010.

But leaving that aside, isn't the name *Troglodytes troglodytes* a wonderful handle for such a little bird? The nesting habits of the bird are contained in a word meaning 'cave-dweller', to reflect a preference for nesting in crevices and holes, or else building a round nest with a side entrance. Human troglodytes still occur in many parts of the world, and I recall seeing some very upmarket cave-houses in the Loire Valley, though one which I saw near Poitiers in 1960 was the home of some junk dealers who lived in total squalor. They could be seen in the city every day, pushing an old pram full of 'unconsidered trifles'.

Wren is derived from a Germanic root which described the cocked-up tail that is so characteristic of the tiny bird. It was this posture which endeared it to ordinary people and resulted in the bird having a number of pet-names, such as Chitty Wren and Puggy Wren. The most widely known of these is Jenny Wren, the sort of homely name that belongs to the days when most people lived a rural life, in close communion with the plants and creatures around them. In folklore, Jenny had a partner, Robin Redbreast, as evidenced in the first lines of the Old Mother Goose rhyme:

Little Jenny Wren fell sick, upon a time.
In came Robin Redbreast and brought her cake and wine... [etc.]

In this, Redbreast was the species and Robin the pet name. Somehow, as bird names were formalised, Jenny kept only her surname while Robin eventually lost his altogether, but not much before the end of the nineteenth century.

However, Redbreast was not noted before about 1400; for at least 400 years before that the bird had been known by another name, Rudduc (a small ruddy-coloured bird). This continued in use for several centuries in the form Ruddock (or Reddock), sometimes with Robin attached, as with Redbreast. It faded from general use during the earlier part of the nineteenth century.

During the nineteenth century, Robin was the most widely used name for the bird, and its popularity was greatly enhanced by its association with the Victorian Christmas, where it became a symbol linked to the nickname of the then red-coated postman. Maybe because of that perceived trivialisation, the name Redbreast was reinstated by the BOU in 1883 and used by Hartert in 1912. As a consequence it appears in Coward in 1926. It was dropped some time afterwards in favour of Robin, which, as Lockwood points out, makes far better sense when used in the naming of 'robins' of other colours.

The colour of the European Robin is represented in its scientific name *Erithacus rubecula,* at least in the 'red breast' of its specific name. The generic name, from the Greek *erithakos*, is of mysterious origins, and long survived in a Latin form as the name of a bird that changed in summer into the Redstart. Today we can readily accept that migratory Robins would move north in spring from their Mediterranean wintering grounds, to be replaced for the summer by Redstarts from further south. However, Aristotle had offered one explanation of the unfathomable seasonal changes in bird populations by suggesting that some species 'transmuted' into other similar species. That idea endured until the eighteenth century, when an understanding of migration started to develop.

EUROPEAN ROBIN

Robin, of course, had by then emigrated, turning up in other parts of the English-speaking world wherever a new species reminded travellers of the bird back home. In that way, a North American thrush, *Turdus migratorius*, a close relative of the European Blackbird, and which just happened to have a reddish breast, became the American Robin. (It is said that members of Avian Equity were scandalised when one made an unauthorised appearance in London in the film *Mary Poppins* alongside Dick van Dyke's pseudo-cockney. However, genuine storm-blown vagrants do turn up from time to time, and one was in London not so long ago.)

Robins of a family not at all closely related to ours, or to the American one, pop up in Australia too: Flame Robin, *Petroica phoenicea*, out-glows them all, as one did for me one misty morning in the Blue Mountains. That generic name means 'rock-dwelling', while the specific name is from the Greek *phoinix* (scarlet). We found the more subtle Rose Robin, *P. rosea*, later the same day. A Scarlet Robin, *P. boodang*, simply stopped the car in Tasmania: the flash of red, black and white by the road was irresistible. (The odd specific name, coined by Gould in 1865, appears to be a corruption of an Aboriginal word, but is otherwise unexplained.) By contrast, Tasmania's endemic Dusky Robin, *Melanodryas vittata*, was as drab as his importance as an island endemic was great. The generic name means 'a dark wood-nymph', though there is not much to justify the specific form, meaning 'banded'. Like that one, the Eastern Yellow Robin, *Eopsaltria australis*, has nothing but the shape to justify the name. Or so I thought until I noticed one eying me boldly on several consecutive days, with the same confidence as the ones in my own garden. It took me a little time to find the nest, so beautifully did it blend with the flaky branches. This bird had lived up to its generic name, meaning 'dawn singer', which had allowed me to locate the territory.

EASTERN YELLOW ROBIN

TIME FOR SOME SERIOUS CONVERSATION: CHATS

The Nightingale and its allies

The voice I hear this passing night was heard
In ancient days by emperor and clown.

Keats's lines remind us just why Nightingale is one of the oldest bird names in the English language. The name has even older Germanic roots, and has barely changed its form since the Old English spelling, *nightegale*. As Lockwood points out, the *–gale* element of the word derives from a word meaning 'to sing'. In that respect, the Nightingale has always held a unique place in European culture, from at least the time of the aforementioned Greek legend of Philomela and Procne. It appears in Christian symbolism; in the medieval works of Chrétien de Troyes and Chaucer; in Milton; in the Romantic poets; and in music. In that latter medium, a Nightingale featured in the BBC's first ever outside broadcast in 1923, when Beatrice Harrison was recorded playing her cello in the garden to the accompaniment of a local Nightingale. The wartime song, 'A Nightingale Sang in Berkeley Square' was a fantasy, not intended to be taken literally, any more than were the Bluebirds over the white cliffs of Dover. The fact that the word appears as a family name, as in Florence Nightingale, suggests that a good voice was behind an original soubriquet which led to the surname: after all, the nineteenth-century Swedish singer Jenny Lind was dubbed the 'Swedish Nightingale'.

The fact that the name now appears as the Common Nightingale must ring with a bitter irony for an older generation who know that it has become so much less common in Britain. I am fortunate enough to live in an area where we can still hear and even see them, since they can often be found by a patient observer during their bouts of day-time singing in April. But it is their night-time singing which makes them special and which is at the root of their great appeal. As a teenager I could often hear one from my room at night, though today Robins chortling in the light of street lamps have to do instead. I recall one magical evening in Poitiers, France, in 1960, when a long walk home after a cinema visit was accompanied by a Nightingale in almost every riverside tree. At moments like that there is nothing commonplace about the Nightingale.

COMMON NIGHTINGALE

The first part of the name *Luscinia megarhynchos* derives from the Latin for Nightingale, and was first applied by Gesner in 1555. It is surprising to find that *Luscinia luscinia* is the Thrush Nightingale, which was described and named by Linnaeus. That species is a summer visitor to Sweden, but the Common Nightingale is not, so it is entirely probable that this was first in Linnaeus's experience. It was not until 1831 that Brehm described and named *megarhynchos*, meaning 'great-billed'. I cannot think that this name was meant to be taken literally, but rather in the spirit of 'loud' (like 'big-mouthed'), meaning 'big-voiced', since it is the better singer of the two.

Another *Luscinia* species is a regular passage species in Britain, the Bluethroat, *Luscinia svecica*. The blue on the throat varies, with some birds carrying a red spot, some a white, leading to the informal reporting of 'red-spotted' and 'white-spotted' Bluethroats. This is down to racial variations, with the nominate race, *L. s. svecica* being red-spotted, and those of southwestern Europe, *L. s. namnetum*, being white-spotted. The third race of the east and south, *L. s. cyanecula*, tends to have white or no spots. The name *svecica* means 'Swedish', which is no great surprise for a bird classified by Linnaeus. The Roman name Namnis, for Nantes in western France, gives the form *namnetum*, while *cyanecula* is a diminutive of *cyaneus* (dark blue).

Siberian Rubythroat, *Calliope calliope*, is a rare vagrant to Britain. Its prowess as a songster is recorded in its scientific name: *Calliope* was the chief of the Greek Muses and renowned for a beautiful voice.

A further closely related vagrant from Siberia is the Red-flanked Bluetail, *Tarsiger cyanurus*. The vernacular name is somewhat laboured, and clearly the sort of latecoming book-name that is frequently found among the American wood-warblers (of which more anon), but it is fully descriptive and is supported by the specific adjective *cyanurus* (blue-tailed). The generic name *Tarsiger* is a little more obscure: it relates to *tarsus* (the lower 'leg' of the bird) and *gerere* (to bear), but the relevance is unclear. I have to say that the one I saw near Bath in 2014 gave me no further help.

In the name of the White-throated Robin, *Irania gutturalis*, the generic form offers a strong clue to its range in Iran and surrounding countries. The specific name simply draws attention to its throat.

The name Rufous Bush Robin, *Cercotrichas galactotes*, appears in the *Collins Bird Guide*, but this name has been replaced elsewhere as Rufous Bush Chat (BOU) and Rufous-tailed Scrub Robin (IOC list). 'Rufous' seems better suited to the nominate race, which appears in Spain in summer, since that is more generally rufous above. In the eastern race, *C. g. syriaca*, which is seen in Greece, only the rump and tail are rufous, which better suits the IOC name. *Cercotrichas* means thrush-tailed, while *galactotes* describes the 'milky-white' underparts of both subspecies.

False starts

My first experience of a Common Redstart was in France, where the similarity to the Robin is underlined by their two names (*Rouge-queue/Rouge-gorge*). The former literally means 'red-tail' and provides a simple clue to the meaning of the suffix –*start*. The word is an old one, with roots in the Germanic forms much older than its first record in English, in 1570. It survived an attempt to impose the name Red Tail, which first appeared in Turner in 1544, but Ray and Pennant both used the form with which we are now familiar.

The Common Redstart is *Phoenicurus phoenicurus*, the name meaning 'crimson-tailed', from the Greek *phoinix* (crimson) and *ouros* (tailed). Crimson is hardly an accurate word for the rufous-orange of the tail. Curiously, in the name of the Black Redstart, *P. ochruros*, we meet the same loose interpretation of colour as in the name of the Green Sandpiper. The Greek word *okros* means pale yellow, yet the actual colour of the tail is not dissimilar to that of the Common Redstart.

Bottoms up!

The Wheatear has nothing whatsoever to do with ears of wheat. The fact that the male Wheatear has a striking black mark through its face may have promoted confusion, but the truth is rather more fundamental in more than one sense. The notable thing about the species is that both sexes exhibit a strong white flash on the rump as they fly away. So the 'white arse' (as rendered in the Anglo-Saxon *hwit aers*) becomes corrupted into the form Wheatear. And, by the way, the word only became 'rude' much later, so there is no point in being coy.

Today the word has become a generic name for a wide variety of related birds: the *Collins Bird Guide* lists seventeen species, of which ours is now the Northern Wheatear, *Oenanthe oenanthe*. Significantly one of these is the Red-rumped Wheatear, while Kurdish and Persian Wheatears also deviate from the name of the genus in having rufous rears. I recall that distant Black Wheatears, *Oenanthe leucura*, showed up on rocky cliffs in Spain only when the white of their tails flashed as they fluttered from one perch to another. This made them instantly distinguishable from the male Blue Rock Thrush, which can look very dark at a distance. The species name *leucura* means 'white-tailed', which would only make sense in a white-tailed group if you had found the feature useful in the way I did in Spain. As if to underline the point, a Greek-based version of 'white-tailed' appears as *O. leucopyga* in the name of White-crowned Wheatear.

NORTHERN WHEATEAR

The generic name *Oenanthe* has a delightful origin. It comes from Aristotle and refers to a bird which coincided with spring – *oenē* means 'vine', and *anthos* is 'bloom' – and presumably also with the grape harvest in autumn.

Today we often hear the term Greenland Wheatear, which refers to *Oenanthe o. leucorhoa*, a race of the Northern Wheatear that tends to be larger, and with distinctive pink-buff undersides. The trinomial is as much use as a chocolate fire-guard, since it just tells us that the bird is white-rumped. This race normally migrates northwards later in spring than the nominate race, and southwards later in autumn, giving two 'waves' of migration for the species.

Among the other European species, the Black-eared Wheatear is the most widespread. Its name, *Oenanthe hispanica*, belies the fact that it is found throughout the Mediterranean region and not Spain alone. That is not the case with the Cyprus Wheatear, *O. cypriaca*, a species which breeds only on that island. It was once considered to be a race of the Pied Wheatear, *O. pleschanka*, which has a far more easterly distribution and a name deriving from the Russian *plesch* (bald spot), a reference to the white head. The bird was described and named by Ivan Lepyokhin, who is known in scientific circles as Lepechin. He was another of the eighteenth-century explorer-naturalists of the reign of Catherine the Great.

Time for a proper chat

Strictly speaking Robin, Nightingale and Wheatear all belong to the group of birds known collectively as 'chats'. Originally the term derived from a sharp call and was applied to other small species too. Today two British species bear that name: Whinchat and Stonechat. The close relationship between the two species is explained by a common ancestry. Their habits exhibit a classic example of 'leap-frog' migration, with European Stonechat migrating only within its breeding range, while Whinchat breeds further north on average and migrates into sub-Saharan Africa. Whinchat has evolved longer wings for that purpose and can often be found migrating with Wheatears.

It seems that their names are intertwined: as Lockwood puts it, 'popular usage frequently fails to distinguish between the two'. Given their habitat preferences – Stonechat is a bird of drier habitats, while Whinchat prefer moister areas – it seems possible that the real value of Whin, an alternative name for gorse or furze, has been tagged onto the wrong species, while an old name, Furze Chat, repeats the apparent confusion.

As for Stonechat, that name underlines the relationships within this group, since the Stone Chack was originally used as a name for Wheatear and makes a lot more sense for the latter, which inhabits stony country and nests in stone clefts.

EUROPEAN STONECHAT

As usual we look to Ray and Pennant for the formal adoption of both these names. In 1678 Ray used Whin-chat and Stone Chatter, and Pennant adopted these ideas in 1786, though using Stone-chatterer. The latter was shortened to Stone Chatt by Tunstall in 1771, and Latham's support for that ensured that the current form eventually became the standard.

Given the odd history behind the name Stonechat, it is ironic that the error is fixed in stone in the generic name of both species, *Saxicola*, meaning 'stone-dweller'.

Early confusion between the two species is also noted by Jobling in the evolution of *S. rubetra* (Whinchat). Aristotle's *batis* was 'a small grub-eating bird', but this was subsumed in a tangle of brambles, which are *batos* in Greek and *rubus* in Latin. The form *rubetra* was used as early as 1476 by Gaza and adopted by Albin in 1731. However, it seems to be an entirely appropriate name to reflect the sort of nesting habitat preferred by this species.

The Stonechat complex has recently been split. Brambles feature more strongly in the new specific name for the European Stonechat, *Saxicola rubicola* (bramble-dweller). The former specific name is now reserved for the sub-Saharan African Stonechat, S. *torquatus*, which means 'collared' and is a reference to the distinctive white neck-patch of the male. It is an echo of the name for Wryneck, but in this case it is not a matter of a twisted neck, but rather a reference to an ancient neck ornament, the torc, which was formed by twisting together several strands of precious metal. The white collar is more extensive in the eastern species, which is now Siberian Stonechat, *S. maurus*. Here the specific name implies 'Moorish' and refers to the darkness of the male's plumage rather than to a geographical range. Ironically, the female is ghostly-pale compared to its European cousin. This change in detail is an excellent example of how quickly nomenclature can change: my 2010 edition of Collins is already heavily annotated in pencil.

THRASHING OUT THE DETAILS: THRUSHES

In Britain there are three resident species of *Turdus* thrushes, plus another which migrates from further south to breed in upland regions. Two further species arrive in good numbers in winter from further north. That odd-looking generic name is simply the Latin for 'thrush', but it is the English word and its related forms which tell a far more complex story.

Lockwood explains that the word 'thrush' has roots which are pre-Germanic. Alongside the eventual development of that word, at least twelve further English dialect and folk-names were formed using the *Thr–* structure. Of those, Throstle and Thrasher are of particular interest.

Throstle was a variant of thrush and the two forms competed with each other, the latter tending to mean Mistle Thrush in regions where the use of Throstle dominated. In other areas, the forms Song Thrush and Mistle Thrush coexisted. Among the ornithologists there was much hesitation, with Pennant leaning towards Throstle, but in 1770 it was White who seems to have ensured that the form Song Thrush was to dominate.

As for the word Thrasher, that form travelled to North America with the settlers, but it is closely similar to Thresher and Thrusher, both of which were used in the Thames Valley and Chilterns areas of England. The thrashers are related to the mockingbirds, but several species are vaguely similar in appearance to Song Thrush and Mistle Thrush. The key visual difference lies in the generic name *Toxostoma*, which means 'arched mouth' (i.e. curve-billed). This feature is not so pronounced in Brown Thrasher, *T. rufum*, which I saw in Florida, but it was still enough to give the bird a distinctively different profile. Incidentally, the bird's red-toned colour is far better conveyed in the Latin name.

The name Song Thrush is well deserved, and is unique among the hot competition for the honour of such a name. The poet Browning pointed to a vital clue for those learning birdsong when he wrote:

That's the wise thrush; he sings each song twice over,
Lest you should think he never could recapture
The first fine careless rapture!

The scientific name *Turdus philomelos* is, perhaps, the greatest of compliments to the prowess of the bird, since *philomela* is the Latin for Nightingale (and returns us yet again to that Greek legend of Philomela and Procne). The strict meaning of *philomela* is 'lover of song'.

SONG THRUSH

The Mistle Thrush gets its name from its habit of eating Mistletoe berries. The name survived an attempt by Merrett and others to introduce the longer form, Mistletoe Bird (or Mistletoe Thrush), which was probably based on the Latin *Turdus viscivorus* (meaning 'mistletoe-eating'). That form was first seen in Aldrovandus in 1599 and is still in use. An alternative spelling, Missel Thrush, was often preferred. Some of our weather-lore has been lost with the decline of the term Storm Cock, which was still in use in my young days: the Mistle Thrush's habit of singing in a tree-top in the face of deteriorating weather was the origin of that term. Lockwood also notes a form which relates indirectly to my earlier comments on the language and social politics of the English Middle Ages (see falcons), where a highly feudal society was ruled over by an often harsh minority of foreign origin. The Mistle Thrush's tyrannical behaviour towards other birds earned it the old name Norman Thrush. In fairness to the bird itself, the openly aggressive behaviour of the species has to do with its unusual nesting practices, since its nests on horizontal bare branches. The species only survives because the parents defend the nest with great vigour to thwart larger predators such as crows. That gets it noticed. I heard quite recently from a very reliable source that he had observed a noisy kerfuffle in which a Mistle Thrush was battling with a Grey Squirrel: the latter was later found dead of its wounds.

The four other regular species do not bear the word 'thrush' in their names. Indeed it comes as a surprise to many people to realise that the garden Blackbird, *Turdus merula*, is a true thrush, even though the most obvious clues are there in the heavy speckling of the young birds. The common name, Blackbird, is in fact far more modern than the roots of the word 'thrush'. Lockwood deems it to be no older than 1300, but of a time when the word 'bird' was still limited to smaller birds ('fowl' being used for larger ones). It is clearly a simple descriptive name.

Naturally enough the modern world requires the name Common Blackbird, since there are other species which now bear the name. Some of those are *Turdus* thrushes, but those of the Americas, such as the Red-winged Blackbird which I saw in Florida and Canada, and the Yellow-hooded Blackbird, which occurs in Trinidad, are icterids (Icteridae), a family which includes grackles, caciques and oropendolas.

Prior to the use of the name Blackbird, the word Ouzel (or Ousel) would have been the norm in Old and Medieval English. This word has roots so ancient that Lockwood shows that they also give rise, by different evolution, to the form *Merula*, the Latin name for the Blackbird, which today serves as the specific name in *Turdus merula*. 'Ouzel' was often used to name other birds of Blackbird size, and most notably Water Ouzel (Dipper). It remains today in the formal name of the migrant Ring Ouzel, *Turdus torquatus*, another black thrush, but this time with a striking white breast-crescent which serves as the 'collar' or 'torc' of the specific name (cf. Stonechat). It is a dramatic and relatively rare bird of the uplands which I have seen only a few times on passage through Berkshire, and only once in the rugged breeding terrain of Northumberland's Cheviot Hills.

Two other *Turdus* thrushes are often known as 'winter thrushes' because they occur in Britain mainly as winter migrants taking refuge from the north. That collective term also strictly includes large numbers of northern Song Thrushes and Blackbirds, which blend with the resident population and are therefore generally overlooked. Strictly, though, the winter Song Thrushes are of the nominate race, while our residents are *T. p. clarkei* (Hartert described the subspecies and named it after fellow ornithologist William Eagle Clarke). However, the most obvious of the winter thrushes are the Redwing and the Fieldfare.

REDWING

The Redwing, *Turdus iliacus*, tends to be more familiar around houses and gardens in winter than its larger cousin the Fieldfare. The wing is red (more accurately a rich rufous tone) only on the under-wing coverts, but the colour extends onto the sides near the wing, so the name is accurate enough. The scientific name *iliacus* refers specifically to the flanks, where the colour is most obvious on the perched bird. The word Redwing first appears in Ray in 1678 and was later formalised by Pennant.

The name Fieldfare is much older. Lockwood shows that a version occurs in about 1100. He suggests that it is too easy to see it as a bird simply 'faring' around fields. He delves into alternative ideas to suggest that the element 'field' has linguistic links to the grey of the bird's plumage, while the *–fare* element seems to link its call to the grunt of a piglet. The bird's names in Welsh and Frisian support this concept. Chaucer's reference to the 'frosty fieldfare' is an interesting link between the grey and the season of its appearance. Not surprisingly the name had many local corruptions and variants, but Ray and Pennant, as so often, were responsible for the acceptance of the current form.

The Fieldfare's scientific name, *Turdus pilaris*, seems to be no simpler, since, according to Jobling, *pilaris* is the Modern Latin for *'thrush'*, but results from an initial confusion between two similar Greek words *trikhas* (a type of thrush) and *thrix, thrikhos* (hair). This leads to an inaccurate Latinising of the word, which appears to mean 'hairless, depilated'. Gesner was the first to record the name in 1555, but others perpetuated the error until it became set in stone by Linnaeus.

The rock thrushes of Southern Europe are closely related to *Turdus* thrushes, but, as 'mountain dwellers', enjoy the generic name *Monticola*. Blue Rock Thrush *is M. solitarius*, which is self-evident, while Rufous-tailed (Common) Rock Thrush is *M. saxatilis* (rock-frequenting), a reflection of its preferences for higher altitudes.

A surprising number of vagrant thrushes are on the British List, originating both from America and from the East (for example Siberia). Such birds are known colloquially among birders as 'Yanks' and 'Sibes' respectively.

Yanks

Several vagrant *Turdus* thrushes occur in Britain, including the American Robin. The genus is widespread, since we encountered three more species in Trinidad. The first was Cocoa Thrush, *T. fumigatus*, whose specific name refers to its 'smoky-brown' plumage. The second was American Bare-eyed Thrush, *T. nudiginus*, whose name literally means 'bare-faced'. The third, White-necked Thrush, *T. albicollis*, repeats its English name in the Latin form.

However, most of the American vagrant species which arrive in Britain, notably Swainson's Thrush, Hermit Thrush, Grey-cheeked Thrush and Veery, belong to a related genus, *Catharus*, which derives from a Greek word *katharos* (pure). That name seems to have been first applied by Bonaparte to the Orange-billed Nightingale-Thrush for its immaculate plumage.

Choate states that the name Veery derives from the bird's call, which Sibley transcribes as 'veer, verre or veeyer'. The name *Catharus fuscescens* (rather dark) seems wholly inappropriate for this rufous-brown bird which Sibley labels as 'reddish overall'.

Of the three other vagrants of that genus, Swainson's Thrush, *Catharus ustulatus*, is named after the English-born naturalist William Swainson, who died in New Zealand in 1855. One of its races is sometimes referred to as the Olive-backed Thrush, though the smoky-grey colour of the adults leads to the specific name *ustulatus* (burnt). Hermit Thrush, according to Choate, was so called for its solitary habits, particularly in winter. It is *C. guttatus* (spotted). We are reminded by the name *C. minimus* that Grey-cheeked is the smallest – and, incidentally, the IOC list uses the English spelling of 'grey'.

Sibes

Five further *Turdus* thrushes occur in Britain as three-star rarities from the east. Black-throated, *T. atrogularis*, and Red-throated, *T. ruficollis*, both have scientific names approximating to the English forms. The white supercilium gives Eyebrowed Thrush its vernacular name, but it is *T. obscurus*, meaning 'dusky'. Confusingly there is also Dusky Thrush, *T. eunomus*, which is quite obscure in meaning 'orderly', but that perhaps refers to the bird's very smart plumage (compare *Catharus*). Naumann's Thrush, *T. naumannii*, is named after the German naturalist Johann Andreas Naumann (1744–1826).

Siberian Thrush belongs to the genus *Geokichla*, which derives from the Greek *geō* (ground) and *kikhlē* (thrush). Its specific name, *sibirica*, is self-evident.

The most dramatic of these thrushes is the largest species, White's Thrush, *Zoothera aurea*. It was named by Latham in honour of Gilbert White of Selborne, the only species to honour the parson-naturalist. The name *Zoothera* means 'animal-hunter', though data quoted in BWPi suggests that it is not likely to hunt prey much larger than a beetle or a grasshopper: instead its feeding habits might give greater justification to the name, since it actively flushes its food items. The specific name *aurea* means 'golden', a reference to the olive-brown upper plumage. It has recently given full species status in a split from the Scaly Thrush, *Z. dauma*, a name of Hindi origin under which White's Thrush still appears in both the BOU list and the *Collins Bird Guide*.

LITTLE BROWN JOBS: WARBLERS

As any birder knows, many of the the 'little brown jobs' which we call warblers are hard to learn and can throw up some real challenges, so there is little wonder that some species of even the quite limited number of British warblers took some time, historically, to sort out, with the late eighteenth century highly significant in the unravelling of the various species. Some of the work, indeed, is still ongoing, since the umbrella term 'warblers' now covers several families which appear less closely related as the scientists re-examine the genetic and other evidence. More noticeable in the second edition of the *Collins Bird Guide* was the appearance of new warbler names, where former subspecies have been upgraded to full species.

The word 'warbler', with reference to birds, was a creation of Pennant in 1773 and was to become the family name of a large number of Eurasian species, as well as a name to denote singing birds of other groups, notably the American wood-warblers.

The OED tells us that the verb 'to warble' is from a much older word, *werbler*, of Norman French introduction, and has therefore changed little in form. Its meaning, 'to sing or play melodiously', is entirely apt for a group which includes some of our best summer songsters – but, arguably, also some of the less musical too. Indeed it is worth considering that without the Blackcap, Garden Warbler and Willow Warbler, the quality of song produced by the remaining British regulars might never have earned the warblers that attractive title. Then what word would the English-speaking world have adopted as a collective name for these birds, or indeed for the wood-warblers of the New World? But those three exist, they sing beautifully, and they have earned the name for all of their allies.

The Old World warblers have now been reclassified into numerous families and several genera, and that is where I will now wander.

Then to Sylvia, let us sing

Well, why not? Shakespeare was extolling a lady of that name, but the *Sylvia* warblers are among our best summer songsters. Gilbert White certainly thought so when he described the song of the Blackcap in 1774 (Letter XL to Pennant):

> ... *when that bird sits calmly and engages in song in earnest, he pours forth very sweet, but inward melody, and expresses a great variety of soft and gentle modulations, superior perhaps to those of any of our warblers, the nightingale excepted.*

The adjective 'sylvan' (from *sylva*, wood) is used, somewhat poetically, to describe a wooded landscape. Scopoli used the word *Sylvia* in 1769 to create a genus meaning 'woodland bird'. Until quite recently that broad meaning still applied to the wider family, the Sylviidae, which included most of the Eurasian warbler species. However, both it and the genus *Sylvia* are names now restricted to a much smaller number of warblers than at the original conception.

In 1915 the BOU decided to consider the Lesser Whitethroat, *Sylvia curruca*, as the type species for the *Sylvia* warblers. The species had been described by Linnaeus in 1758, but because of confusion over the accuracy of his notes on what may or may not have been Common Whitethroat, *Sylvia communis*, the latter species is now attributed to Latham in 1787.

LESSER WHITETHROAT

In 1768 Gilbert White was the first in Britain to separate the Lesser Whitethroat. The current name was created by Latham in 1787. The Latin word *curruca* was for an unidentified small bird.

The Common Whitethroat has traditionally been so common (a fact which is reflected in *communis*) that the simple name Whitethroat has normally been sufficient. It was first in evidence in 1676, but Lockwood suggests that it is much older, as a simple descriptive name like Redbreast.

In like vein, the name Blackcap is also a form with a long history, though traditionally applied to almost any species with a black head, among others Reed Bunting, Marsh Tit and even Black-headed Gull. Pennant was the first to apply it specifically to *Sylvia atricapilla*, though some subsequent naturalists used Blackcap Warbler, presumably in the light of the earlier breadth of usage. The Latin form, *atricapilla*, means literally 'black-haired', which is near enough.

GARDEN WARBLER

The scientific name of the Garden Warbler, *Sylvia borin*, is the butt of an irony which suggests that the g is missing. This calumny is entirely down to the fact that the best way to describe the appearance of the bird is to say that it has no outstanding features. I read somewhere a description which suggested that the bird is in the 'biscuit porcelain phase' – unpainted and unglazed – which I feel is much more apt for a bird which should be praised for its song rather than done down for its dowdy suit. The Nightingale suffers no such insults, yet it is too has plain plumage suitable for a skulking bird. The word *borin* is a curious one: it is an Italian Genoese word associated with a species which keeps company with cattle – from the Latin *bos* (ox). I suspect that this would more likely be the Yellow Wagtail, which is a more subdued colour in winter plumage, rather than the arboreal warbler. The name was given to the warbler in 1783 by the Dutch naturalist Pieter Boddaert, who was cataloguing a friend's collection of specimens, and was probably selected with little awareness of the bird's actual habits. The vernacular name Garden Warbler also seems rather inaccurate, certainly in the context of the average garden. It was first applied by Bewick in 1832 and taken up subsequently by Gould and Yarrell, but the origin of Bewick's use appears to be Gmelin's *Sylvia hortensis* (Latin *hortus* for garden), which turned out to be an Orphean Warbler (now Western Orphean Warbler). All in all, the naming of the Garden Warbler leaves a lot to be desired.

Given all of the above, it is worth stepping back to what went before, since the Garden Warbler had been known to earlier ornithologists by the delightful name of Pettichaps (probably meaning 'small fellow') a dialect word used by Ray in 1678, with Greater Pettichaps and Lesser Pettichaps distinguishing the Garden Warbler from the other brown and 'featureless' warblers, Chiffchaff/Willow Warbler (see later).

There is only one other reliable representative of the genus in Britain, the Dartford Warbler, *Sylvia undata*, which, in Britain, lives at the extreme north of its range as a rather precarious resident, extremely vulnerable to bad winters. The origins of the common name of this bird relate to those of Sandwich Tern and Kentish Plover, since shot specimens were sent to Latham from Dartford in Kent. This time he sent them on to Pennant, who standardised the name Dartford Warbler in 1776, though the formal scientific description and naming is again attributed to Boddaert, who in 1783 highlighted the distinct throat-speckling of the male in the name *undata*.

A trip to Andalucia produced a very interesting new bird name, since birds came so thick and fast in a marsh at Malaga that we found ourselves calling new sightings in shorthand. In that way Sardinian Warblers became *Sardines!* Their specific name, *Sylvia melanocephala*, makes them the true 'black-headed' Sylvia species, but this time in Greek. The Subalpine Warbler, which we saw elsewhere at a suitable elevation to support its English name, might well be called the Warbling Warbler if we took its specific name *S. cantillans* quite literally. A recent split has created a new species, Moltoni's Warbler, which is confusingly known as *S. subalpina*.

This genus is well-represented by other species in Europe, several of which have names which bring out salient features.

Barred Warbler, *Sylvia nisoria*, has a specific name which indirectly echoes the English name, since the barring on the breast recalls that of the Sparrowhawk, *Accipiter nisus*. That is hopefully rather a better explanation than any propensity to become the hawk's dinner. I was recently watching my first ever Barred Warbler, which had stopped to feed on passage: some extra magic was added to that moment by the sight of a Sparrowhawk soaring nearby.

Eastern Orphean Warbler is *Sylvia crassirostris*, so called for its 'strong bill'. The name Orphean is a reminder of Orpheus, the Thracian minstrel of Greek myth, and links to its clear, sweet song. Cyprus Warbler is *S. melanothorax*, a name which describes the dark-grey ground-colour of the breast. Spectacled Warbler, *S. conspicillata*, has names which emphasise the distinct white eye-ring in both English and Latin.

Two more species record the names of people. Ruppell's Warbler, *Sylvia rupelli* (recently amended from *rueppelli*), honours the German collector, Peter Rüppell (1794–1884), who has seven other species to his name. A Napoleonic general and sometime naturalist is commemorated in the Marmora's Warbler, *S. sarda*, while the specific name means 'a Sardinian woman'.

In a class of its own

As a relative newcomer, Cetti's Warbler, *Cettia cetti*, is probably the least well known of British warblers. It is in any case scarce and local, and is a notorious skulker. Its song seems to call its wonderfully alliterative name, but that was in fact coined by Temminck to honour the eighteenth-century Italian Jesuit priest and naturalist, Francisco Cetti, who had observed the species in Sicily. It is the sole European representative of a family which is otherwise well represented in Asia and Africa. As a resident in Britain it is at the northern extreme of its range and is therefore very susceptible to harsh winters.

Where's the point?

The *Acrocephalus* warblers might readily be described collectively as 'marsh or reed' warblers, but as these are also specific names, such a term is best avoided. It is a pity that there is not an alternative as neat as *Sylvia*, since the present term is somewhat clumsy. It was a creation of father and son, J. A. and J. F. Naumann, in 1811. The BOU, in 1915, suggested that their interpretation of the Greek *akros* (topmost) was in error, since the Naumanns had intended a meaning closer to the Latin *acutus* (sharp-pointed) to describe the attenuated shape of the head (*kephalē*).

There are in fact some thirty-five or so species, of which the two most common in Britain are Reed Warbler and Sedge Warbler. The subtle difference in the English names reflects the fact that, while the two use similar habitats, the latter has somewhat more liberal habits. Both names are traditional in their roots: Reed Wren and Sedge Bird were both in use before the ornithologists used them, and it was Pennant who harmonised the two as Reed Warbler and Sedge Warbler. However the former was not to become standard until Yarrell in 1843, because Latham, Selby and Gould all preferred Reed Wren.

The specific name of Reed Warbler, *Acrocephalus scirpaceus*, could not be simpler: it derives from the Latin *scirpus* (rush or reed). The case of the Sedge Warbler, *A. schoenobaenus*, is a little more complex, since it appears to be a translation of the old Swedish name *Savstigare* (meaning 'reed-climber' or 'reed-treader') into Latinised Greek from *skhionos* (reed) and *bainō* (tread). Linnaeus was responsible for that, of course.

REED WARBLER

Though these two relatively common species light up many a British reed-bed in early summer with their quirky songs, others of their genus are not so easily found. However, I doubled my British list of the *Acrocephalus* warblers in 2012 with Paddyfield Warbler in late winter, and a Marsh Warbler in mid-June. In birding, such unexpected opportunities give the essence of the activity, and being a birdwatcher rather than a list-ticker, I went back to see each a second time in order to learn a bit more about their style and habits.

Closely resembling the Reed Warbler, the Marsh Warbler, *Acrocephalus palustris*, was formally separated in Britain in 1871, but had been described and named by Bechstein in 1798. The common name is a translation of the scientific name. There has always been a small breeding population in England, though the bird is elusive and very scarce these days. My sighting was of a bird singing in a patch of waterside scrubby herbage which is typical of the species.

My Paddyfield Warbler, *Acrocephalus agricola*, was something of a celebrated oddity, since it wintered at Pagham Harbour in West Sussex and obligingly stuck to a defined area of reeds, where I saw it on two visits. It was a true vagrant, having travelled completely the wrong way on a migration which should have taken it from the Balkans or Central Asia down to the Indian subcontinent. As its name suggests, it is associated with rice as well as reeds: 'paddyfield' is the name given to areas where rice (*padi* in Malay) is grown. The specific name *agricola* means 'field-dweller', and was given by Jerdon in 1845 – a rather lame choice in the circumstances, given the number of species which might earn such a name.

When I first saw Australian Reed Warbler, *Acrocephalus australis*, it was still classified as a subspecies of the Asian Clamorous Reed Warbler, *A. stentoreus*. On the second visit it was a full species and my lists needed to be changed. Clamorous or not, it proved to be a sturdy singer, though manifestly less shy than our own (European) Reed Warbler. *Stentoreus*, of course, means much the same as 'clamorous'.

Back in Europe there are four other *Acrocephalus* warblers. Aquatic warbler, *A. paludicola*, is a 'marsh-dweller', so we have little doubt with those two names telling of its habitat. Such matters feature also in the names of the Great Reed Warbler, *A. arundinaceus* (reed-like), and Blyth's Reed Warbler, *A. dumetorum* (of the thickets). The nineteenth-century zoologist Edward Blyth appears in the name of this species: he is also recorded in the names of an astonishing fifteen further species. His portrait shows him with a superb set of facial hair, a feature which is recorded in contradictory ways in the names of the Moustached Warbler, since its name *A. melanopogon*, means 'black-bearded'.

Coining a name or two

A further genus, the *Phylloscopus* warblers, is well represented in Britain. The name means 'leaf-searching', since most spend time gleaning their food items from the leaves and twigs. The genus was first proposed by Boie in 1826 and forms part a group of 66 species known as 'leaf warblers'.

It is a well-told tale that Gilbert White was the first in Britain to conclude from his daily observations that 'there are three species of willow-wren'. His descriptions leave no doubt that he had observed the birds closely, as in this account of what we today call the Wood Warbler:

> *The yellowest bird... haunts only the tops of trees in the high beechen woods, and makes a sibilous grasshopper-like noise, now and then, at short intervals, shivering a little with its wings when it sings. (Letter XIX to Pennant)*

With the decline in the numbers of this species, fewer people get to hear and see that behaviour now, but I well remember that my first attempt to photograph one in the New Forest produced a wonderful blur as the bird did the shivering part. This is *Phylloscopus sibilatrix*. In 1793 Bechstein echoed White's use of 'sibilous' to give the bird a name which means 'whistler'.

White was the first to use what was probably a local name to refer to the Chiffchaff. Yarrell used Willow Warbler and Wood Warbler in 1843, to bring them into line with the systematic pattern, but still respecting the earlier names, Willow Wren and Wood Wren.

The word 'wren' still remains in the name *Phylloscopus trochilus* (Willow Warbler), since that is the meaning of the Greek *trokhilos*. Linnaeus named the species in 1758. An echo of an echo remains in the name of a species which is a vagrant to the British Isles, the Greenish Warbler, *P. trochiloides*, since that means 'resembling *trochilus*'. The Two-barred Warbler, *P. plumbeitarsus* (with lead-coloured legs) has recently been split from the former and renamed.

The specific name of the Chiffchaff, *Phylloscopus collybita*, comes from the Latin *collybista*, which means 'money changer'. It was coined by Vieillot in 1817, to echo a popular name, *compteur d'argent*, which was used in his native Normandy. It is a rather graphic rendition of the bird's repetitive song as the sound of chinking coins.

CHIFFCHAFF

In recent years there have been considerable changes made within the genus, with the split of Iberian Chiffchaff, *Phylloscopus ibericus*, and others, which now necessitates the international name Common Chiffchaff for the species which we see in Britain. However, Siberian Chiffchaff, which has been confirmed as a British winter visitor, is still considered a race of the Common Chiffchaff, as *P. c. tristis*. The trinomen means 'sad', and reflects the mournful tone of the call, as well, perhaps, as the drab greyish plumage. That grey plumage is not dissimilar to the Scandinavian race, *P. c. abietinus*, which is another winter visitor to Britain. The trinomen here means 'of the fir trees'. It is interesting to note that both

Siberian Chiffchaff and Scandinavian Chiffchaff were listed as a subspecies by Coward in 1926, but the Iberian form did not appear, presumably as none had been identified in Britain at that stage.

A further species has so increased its presence in Britain in recent years that it is now classified in Collins only as a one-star vagrant. Part of the population of the Yellow-browed Warbler, *Phylloscopus inornatus*, according to one theory, may have changed its migratory patterns to now travel in greater numbers through Europe. In spite of its English name focusing on a distinctive feature, *inornatus* means 'plain'. That is down to Blyth in 1842.

Other members of this genus include Arctic Warbler, *Phylloscopus borealis*, which means 'northern', and Dusky Warbler, *P. fuscatus*, which is 'dusky' in both languages. The rather similar Radde's Warbler, *P. schwarzi*, has one of those rare names which commemorates two people: in this case, both were German scientists who took part in the East Siberian Expedition of 1855. Gustave Radde found and named the species, which bears both his name and that of the expedition leader Ludwig Schwarz. Radde later had the honour of becoming the chairman of the first IOC Congress in 1884.

Several other naturalists are also recorded in this genus. Western Bonelli's Warbler, *Phylloscopus bonelli*, and Eastern Bonelli's Warbler, *P. orientalis*, are both named after the Italian naturalist, Franco Bonelli, who is also commemorated in the eagle. Hume's Leaf Warbler, *P. humei*, commemorates the British-born Allan Hume (1829–1912), who wrote widely on Indian birds, and whose name appears in more than a dozen other species. Peter Pallas appears yet again in Pallas's Warbler, *P. proregulus*, which also reminds us that it is extremely similar in size and appearance to the 'crests' of the genus *Regulus*.

Reeling them in

The next warbler genus represented in Britain is *Locustella*, a term created by Kaup in 1829. It is a diminutive form of the Latin *locusta* (grasshopper), which of course appears in the word 'locust'. Naturally enough, the type species for this genus is the Grasshopper Warbler, *Locustella naevia*, which gets its English name from that frustratingly hidden reeling song delivered from somewhere deep in the scrub. The name grew from a phrase used by Ray in 1678 to describe the 'Titlark that sings like a Grasshopper'. This was taken up by Pennant, first as Grasshopper Lark, and later as Grasshopper Lark Warbler, but Latham used the current form in 1783. Now we strictly need to call it the Common Grasshopper Warbler to separate it from Pallas's Grasshopper Warbler and others.

As for *naevia* (Boddaert, 1793), the word means 'spotted', and indeed the spotting of the under-tail coverts and mantle is strong. Such features are totally absent in the one other scarce British breeding member of the genus, the Savi's Warbler, *Locustella luscinioides*. The nineteenth-century Italian naturalist Paolo Savi is commemorated in that species: Savi named and described it in 1824 and Yarrell coined the English name in 1843. The scientific name, *luscinioides*, means 'Nightingale-like', but it seems to fit in no way: the song is a monotonous, mechanical, insect-like reeling, so unlike that of the Nightingale, while physical resemblance between the two species is tenuous to say the least.

Three members of this genus earn a place as rare vagrants to Britain from

Siberia – 'Sibes' in the jargon. Spotting is most pronounced in the Lanceolated Warbler, *Locustella lanceolata*. The names are explained by the bold 'spear-shaped' spots on the breast. The German Peter Pallas is again commemorated in the name of Pallas's Grasshopper Warbler. Its scientific name, *L. certhiola*, suggests that it is 'like a little treecreeper' – and the two are in fact of similar size. The third species is River Warbler, which repeats the concept of its English name in *L. fluviatilis*.

Look what crawled out from under a stone!

In 1827 Conrad von Baldenstein, a Swiss naturalist, applied the name *Hippolais* to another major genus of warblers. This genus is represented in Britain only by a few vagrants. *Hippolais* is a word of disputed origins, first used by Aristotle and others as the Greek *hupolais* and probably, according to some, referring to the Northern Wheatear. If that is the case, the BOU's 1915 explanation is highly acceptable, in that it is the name of a bird which creeps under stones (*hupo* is 'under' and *laas* is 'stone'). In addition, the BOU stated that Linnaeus had rendered it inaccurately. As with so many of these Ancient Greek birds, the name was recycled in an arbitrary way, leaving this 'tree warbler' (a vague term of convenience for this genus) tainted with a piece of wholly uncharacteristic behaviour.

Jobling refers to Jonsson's interpretation of the call of Icterine Warbler as *hippolüeet*, but since Svensson renders it as *teh-teh-lüüit* and BWPi illustrates five more forms which all differ as widely, it seems safe to assume that the association of the call with the genus name is but a red herring.

Icterine Warbler, *Hippolais icterina*, has names which refer to its colour as 'jaundice-yellow', which reminds us that the sight of a Golden Oriole was considered in Greek superstition to be a cure for jaundice.

I had to wait until quite recently to see the very similar Melodious Warbler, *Hippolais polyglotta*, during a trip to Spain. As the English name suggests, much of the song is pleasant, though some passages are harsh. Like the Marsh Warbler it is a great imitator, which fact earns it the specific name *polyglotta* (which can mean both 'multilingual' and 'loud-voiced') – and which, confusingly, also occurs in the Spanish name for the Marsh Warbler.

There is little mystery in the names of the Olive-tree Warbler, *Hippolais olivetorum*.

Suitable, or not?

Europe has yet another genus, the *Iduna* warblers, which appear in the *Collins Bird Guide* under *Hippolais*.

This re-designation seems to be a reversion to a nineteenth-century form, of which the meaning is obscure. Jobling offers no fewer than three possible roots: the Latin *idoneus* (appropriate or suitable), the Greek *idou* (behold), or a literal spelling of *aedonis* (Nightingale). It is much harder to imagine the relevance of the first two than of the third explanation.

The generic name is not the only alteration evidenced in the IOC list, where Olivaceous Warbler is now prefaced by the word 'Eastern', to distinguish it from the Western Olivaceous Warbler, which is now favoured over the name Isabelline Warbler, so out go Queen Isabella's grimy underclothes! The former is *Iduna pallida*, which the Collins guide aptly describes as 'like a washed-out, greyish Reed

Warbler'. The latter is *I. opaca* (dark), which is an odd way of describing a bird labelled as 'sandy-brown' in that same reference. Booted Warbler, *I. caligata*, has less contradictory names, since the specific name means the same as the English. It is so called because the toes are darker than the tarsus. *Caligata* recalls the infamous Roman Emperor Gaius, who is better known by the nickname awarded to him as a child by his father's troops when he appeared in miniature armour: Caligula meant 'Little Boots'. Sykes's Warbler is a relatively recent split from Booted Warbler, and one which commemorates Colonel William Sykes of the Bombay army, whose work as a statistician led him to the task of cataloguing Indian birds. As *Iduna rama* that species is named for a reincarnation of the Hindu god, Vishnu.

The fan club

A bird which is rare in Britain is still known as the Fan-tailed Warbler on the BOU list. The bird's habit of spreading its tail wide during flights accounts for 'Fan-tailed'. Internationally, however, it now has an entirely different name, Zitting Cisticola, which has to be one of the most bizarre of all bird names – a fact which perhaps lends it a certain cachet. The BOU clings to the traditional form for no clear reason – apart, perhaps, from tradition or sentiment.

The cisticolas are a very widespread family, now considered to be quite closely linked the grasshopper warblers, with species in Africa, Asia and Australia. Cisticola derives from the Greek *kisthos, kistos* (rock-rose) and *–cola* (dweller), which indicates a habitat type. The word 'Zitting' makes little sense at all unless one has heard the flight call: 'zit, zit, zit!' Some thirteen species of the family have equally bizarre names, such as Wailing, Churring and Tinkling Cisticola, all based upon the voice of the species. The significance of that is that it is almost impossible to distinguish them visually, so the voice becomes the major means of identifying them. The specific name of Zitting Cisticola is *Cisticola juncidis*, from the Latin *iuncus*, which indicates that its habitat is reed-beds. The same word is at the root of the name of the American sparrow, Junco. In Australia we encountered the more soberly named Golden-headed Cisticola, *C. exilis*, the specific name meaning 'slight or slender'. It too was a dweller of reed-beds, this time in a marsh fringing the Yellow Waters in Kakadu.

Crests and crowns

The small number of birds of the genus *Regulus* was originally classified by Latham with the warblers, though the celebrated French palaeontologist Georges Cuvier (1769–1832) later separated them. They now have been placed much closer to the wrens, which links them back to one of the earlier name-forms used in Britain until the early twentieth century.

In Britain the two members of the family are Goldcrest and Firecrest, these names leading to the popular collective term 'crests'. In North America, where there are also two species, these are known as kinglets (Golden-crowned and Ruby-crowned). The fact that they are all 'crested' or 'crowned' is the origin of the name *Regulus* (little king/kinglet). The colours implied in the vernacular names are all variants on yellow-orange to scarlet and are strikingly exposed only when the bird is agitated or displaying.

It was Ray who first referred to the Golden-crown'd Wren, which Pennant changed to Golden-crested Wren. Uncharacteristically, Yarrell used the whole arsenal of Little Golden-crested Regulus or Kinglet. The form Gold-crest asserted itself: it was used by the BOU in 1883 and survived Hartert's re-use of Golden-crested Wren in 1912.The hyphen had disappeared by 1926 in Coward's version of the 1923 BOU list.

The Firecrest had a later start, since it was not a breeding species in Britain until fifty years ago. The pattern of the name, coined by Eyton from Temminck's Latin, *ignicapilla*, of 1820, largely tracked that of Goldcrest to its current form.

The scientific name of the Goldcrest, *Regulus regulus*, is rather uninteresting, but the Firecrest has a little more flare, one might say, with the Latin form *R. ignicapilla* telling him that his hair is on fire! The Ruby-crowned Kinglet is *R. calendula* – just 'glowing', but reminding us of the colour of the marigold, also *Calendula*. The Golden-crowned Kinglet is *R. satrapa* from the Greek name *satrap* for a Persian viceroy – a delightfully oblique reference to a crowned head. That is the one that I have yet to see.

SPOT THE FIG: FLYCATCHERS

Pennant modelled the word 'flycatcher' exactly on the Latin, *Muscicapa*, providing also the two species names, Spotted and Pied. 'Flycatcher' is a simple way of describing a form of hunting action in which a bird sits on a perch, spots a passing insect, and then flies out to catch it. Since I have seen such species as Chaffinch and Willow Warbler sometimes hunt in this way, the action alone is not a guarantee that a bird observed in silhouette is necessarily either Spotted or Pied Flycatcher, but that action is often their giveaway.

SPOTTED FLYCATCHER

The Spotted Flycatcher was well known and had a number of popular names, several of which relate to the typical nesting site of the species: Beam Bird, Post Bird and Wall Bird. I well recall a Spotted Flycatcher which nested annually during the 1960s in the wisteria growing on the wall of a cottage, but then their numbers declined and the birds failed to return. The scientific name, *Muscicapa striata*, contains a rather more accurate description of the bird's 'streaked' markings than the word 'spotted'.

Pied Flycatcher apparently had no previous vernacular names in English, though it has local names in Welsh, a reflection of its bias towards the west. The scientific name, *Ficedula hypoleuca*, is a curious one. The word *Ficedula* means 'fig-eating'. A sortie into the text of the BWP confirms that, though insects and the like make up most of its diet, the Pied Flycatcher also takes some berries and fruit, including figs. In Ancient Greece the bird was thought to change into a Blackcap in winter, this by a process known as 'transmutation'. The concept was considered by Aristotle to explain the then incomprehensible phenomenon of migration and was perpetuated as a theory until the eighteenth century. As for the specific name, *hypoleuca*, it is something of an enigma, since it means 'less than white' or 'whitish'. This suggests that Pallas was considering the subdued non-breeding male or the female, rather than the strongly pied breeding male, when he named it in 1764.

Of the other related European species, Collared Flycatcher and Semi-collared Flycatcher have scientific names which echo the English: the former is *Ficedula albicollis* (white-necked) and the latter *F. semitorquata* (semi-collared). However, Red-breasted Flycatcher is *F. parva* (small), and Taiga Flycatcher is *F. albicilla* (white-tailed).

The flycatchers of the Americas belong to an entirely different family, the Tyrannidae. In Trinidad, some members of the family were a real test of our patience: little shadowy birds populated shadowy branches, and in truth I would have identified very few without expert local guidance. Fortunately some, like the Greater Kiskadee, were good obvious targets. I have found a few phoebes and kingbirds in North America, though the challenge of identifying a tiny Yellow-bellied Flycatcher in a Cape Breton wood was one of my more successful moments with this family, thanks partly to a short video shot of it.

Also unrelated are the Australian flycatchers, the most obvious of which is the Magpie-Lark. Three less obvious species became familiar during our trip to Kakadu. These all belong to the Monarch family.

(Typo flycatchers also exist in my world. Potted Flycatcher is one bottled for the Mediterranean cuisine. In that context, Pied Flycatcher comes with a crust.)

ON A TREE BY A RIVER... TITS AND CHICKADEES

A bit of Gilbert and Sullivan this time.

Lockwood explains that while the origin of the word 'tit', in the context of birds, is old, its current acceptance as the definitive word is relatively modern. It first appeared as a bird noun in 1706 as an alternative to the much older form 'titmouse', which was in evidence as early as 1325 in the spelling *titemose*. The first element, *tit–*, was originally used adjectivally in the sense of 'small'. The OED suggests a Scandinavian origin for that. The similarity of the second element, –*mose*, the bird's name, to 'mouse' led to the word following the pattern of mouse, including the irregular plural, in the form *titmice*. Both the shorter and longer forms were subsequently used by ornithologists, with Yarrell preferring the shortened version in 1843, while the BOU in 1883, Hartert *et al.* in 1912, Coward in 1926, and even the *Checklist* in 1952 all retained use of the longer form. It is entirely possible that something more than ornithology held back the change: Macdonald quotes the fact that 'a curious Victorian prudery' caused about forty name changes on the Australian list in 1926, when terms like 'rump', 'belly', 'vent' and 'breast' were the offending words. Even quite recently, David Lindo reported being rebuked in an American hide for inadvertently referring to chickadees by the shorter British name, which had become the norm during the second half of the twentieth century.

The Gilbert and Sullivan Tom-tit quoted above was a local name for the Blue Tit (and sometimes others) in Britain, but was also applied in America to the Tufted Titmouse as Tomtit. The common usage suggests that it was a very old form, and it has an obvious link to the convention of attaching familiar names to familiar birds, as in Robin Redbreast and Jenny Wren.

Having retained the longer form in such names as Tufted Titmouse, the Americans found a convenient alternative in the song of the commonest species, the Black-capped Chickadee, whose three-note call provided an onomatopoeia which served well as a generic name for their other species.

As for the current specific names of the British species, there was a relatively consistent pattern of development: in 1544 Turner used Great Titmouse; Merrett noted the Long-tail'd Titmouse in 1667; Crested Titmouse appeared in Charleton in 1668; in 1678 Ray referred to the Blue Titmouse and the Marsh Titmouse; and in 1771 Tunstall used Cole Titmouse. In every case it was Yarrell who imposed these names in a standardised way, in 1843, as Great Tit and so on, though still with the spelling Cole Tit, which was not fully modernised to Coal Tit until the end of the nineteenth century.

There are elements of folk-names in some of these, as in Blue Cap and Coalmouse, both of which focus on the colour of the cap, but generally the shapes of all these names were guided and chosen by ornithologists. In the case of the Great Tit, for example, it was a straight translation of the Latin form, *Parus major*. One of the more interesting of its prior names is Oxeye, which comes from a Norman French term, *oeil de boeuf*, for a small bird (and also applies to the small centre target in archery, the bull's-eye). As for Marsh Tit (one of the birds originally known as Blackcap), the name was again a straight translation of Gesner's erroneous use of *palustris* (marshy) in 1555, and nobody seems to have bothered to challenge its validity for a woodland species. Long-tailed Tit is another 'book-name', as Lockwood puts it: this leaves some very attractive and imaginative folk-names strewn in its wake – names as varied as Kitty Longtail, Bottle Tit and Feather Poke ('poke' being a bag, and referring here to its nest). Crested Tit, as a species of remote areas of Scotland, had no English folk-names. The odd man out is the Willow Tit, which was not identified as a British species until 1897 and was named for a tree which it does frequent (being more of a marsh tit than its cousin).

GREAT TIT

Yarrell's objective was to use a standard pattern of names for this family, which were at the time all classified as *Parus* (meaning 'tit'). With the exception of the Long-tailed Tit that situation still pertained in Coward's 1926 list. Today things are very different, with only *Parus major* surviving a reshuffle.

Coal Tit is now *Periparus ater*, which seems to suggest that it is 'close to the *Parus* genus'. The specific form *ater* (dull black) is less woolly in its intention. (In the world of typo birds the Coal Tit is very closely related to the Grate Tit, by the way.)

The Blue Tit is now bluer than ever, as *Cyanistes caeruleus*, blue in both Greek and Latin. The continental Azure Tit follows that example as *Cyanistes cyanus*, which underlines the fact that is a different shade of blue.

Marsh Tit, *Poecile palustris*, and Willow *Tit*, *P. montanus*, belong to a genus which is based on yet another unknown Greek bird, *poikilis*. Needless to say, *P. montanus* has a much wider range of altitude preferences than the suggested 'mountains'. The Siberian Tit is *P. cinctus*, meaning 'banded', a name which comes from a description by de Buffon of it as a 'white-banded tit', though it is in fact only white-cheeked. For some unfathomable reason, given that only one small part of the Siberian Tit's range is in North America, the American name, Grey-headed Chickadee, now prevails in the IOC list. Sombre Tit is *P. lugubris*, which means much the same as the English.

The Crested Tit is now *Lophophanes cristatus*. The generic name derives from Greek to state that it is 'showing a crest'. The Latin *cristatus* tells us that it is 'crested'. As I have said before, overstatement is not uncommon in the world of scientific names.

The Penduline Tit, *Remiz pendulinus*, makes occasional appearances as a vagrant in Britain and I have narrowly missed the chance to see them on more than one occasion. The 'penduline' element of the name refers to the hanging nest which the bird constructs, while the genus *Remiz* is the Polish name for the species. This genus is closely related to the other tits.

As *Aegithalos caudatus*, the Long-tailed Tit is now endowed with a generic name used by Aristotle for several of the tit species. In spite of that it is now classified in a different family, the bushtits. Curiously the specific name, *caudatus*, merely draws attention to the tail, without reference to its length.

LONG-TAILED TITS

PERPETUUM MOBILE: BEARDED REEDLING

'Perpetuum mobile' is a musical form which is always on the go, and is perhaps best known in Rimsky-Korsakov's Flight of the Bumblebee. It seems an apt metaphor for the constantly changing names of this species.

At the time of Yarrell's attempt to bring order to the supposed *Parus* genus, one other species was included, which has now proved to be more closely related to the larks than to the tits. The modern usage of the name 'Bearded Tit' begins with Brisson in 1760, who called it *Parus barbatus,* a translation of the French *mésange barbue*. That form was adopted and Anglicised by Tunstall in 1771 as the Bearded Titmouse. Even today, the BOU retains the name Bearded Tit, possibly because it is so much simpler and less clumsy than Bearded Reedling, the formal international name. At least Reedling is an old Fenland name for the species and is probably older than the form imported from Brisson. The hybrid name is too much of a mouthful and seems to have little merit, since 'bearded' is, in any case, an odd way to describe a species in which the male has two dark moustachial stripes. Many birders call them Beardies, and might convert to Reedling, since it falls in with Dunnock as a simple folk-name, but the acceptance of Bearded Reedling, I predict, will be very slow indeed.

BEARDED TITS

The scientific name, *Panurus biarmicus,* has now placed the species in a genus of which it is the unique member. *Panurus* simply tells us in Greek that it is 'long-tailed'. The word *biarmicus* is much more complicated, since this creation of Linnaeus was misinterpreted by Cuvier and Temminck to mean 'bearded', whereas Linnaeus himself had intended it to refer to a district of Russia named Bjarmland (Latinised as Biarmia), which was an ancient state on the southern shores of the White Sea. The reasons for that connection are incomprehensible, since the area is a long way north of the species' range. In a way, Cuvier and Temminck's interpretation was a happy side-track. If ever there was a case to call 'pass the eraser' this would seem to be it (indeed, *Passer aser* would be a good name for any bird of dubious status), but in the 'archaeology' of birds' names that would be the equivalent of blaming an Etruscan for not inventing gunpowder. This confusing name, *biarmicus,* is a good example of that 'oldest fool' rule, where a flawed name survives in spite of its inaccuracy. This seems to be a good point to mention that Linnaeus was not primarily an ornithologist: birds were only one relatively small part of his immense work on the classification of *all* living things, so he may readily be forgiven the occasional slip, eccentricity or inaccuracy in the details of a Herculean task. My job is to unearth the stories, but my intention is certainly not to cast the first stone against a man who towers as a giant in this story.

HAMMERERS AND PLASTERERS: NUTHATCHES

I earlier discussed how the development of the word 'titmouse' illustrated the confusion of two words. Lockwood explains that this happens again in the name of the Nuthatch, which has its origins in such as *notehache, nuthak* and *nothagge,* all of which are related to the word 'hack'. It was the similarity of 'hack' and 'hatch' which led to confusion and to the present form. The alternative name used by Ray was Nutjobber, where 'job' is akin to 'jab' in its meaning. Strictly, therefore, both names followed the observation that the bird hammers nuts and acorns to open them. The form we now use therefore has roots which probably stretch right back to Old English. It was Pennant who preferred it over Ray s alternative, and who effectively standardised it.

Lockwood also notes the old names Mud Dauber and Mud Stopper, which record another unusual habit of the bird in reducing the nest-hole with mud to a size which suits it. That occurs not only in trees, but also in the rocky crevices used by the rock nuthatches. So far as I can ascertain, there was no name based on the bird's unique ability, among British birds at least, to run head-first down a tree-trunk.

EUROPEAN NUTHATCH

As is so often the case, the name of one British species has now become the English name for a whole genus, making ours the Eurasian Nuthatch, *Sitta europaea*. I have, on occasions, watched the Red-breasted Nuthatch, *S. canadensis*, in Nova Scotia. *Sitta* derives from a Greek word *sittē*, which was used by Aristotle and others to refer to a small, woodpecker-like bird.

Two European species, the endemic Corsican Nuthatch, *Sitta whiteheadi*, and the Western Rock Nuthatch, *S. neumayer*, respectively recall two naturalists, the Englishman, John Whitehead (1860–1899), and the Austrian, Franz Neumayer, who died in 1840.

SOME CLIMB UP TREES AND BARK... TREECREEPERS AND WALLCREEPER

The phrase above comes from an old un-punctuated riddle, which begins: 'Time flies you can't they fly too fast.'

The evolution of the name Treecreeper begins with a Latinised version, *creperam*, in Turner in 1544. Creeper is used by both Ray, in 1678, and Pennant, in 1768. Albin, meanwhile, had coined the form Common Creeper, which was used by Yarrell in 1843. The modern word Treecreeper first appeared as a two-word name, Tree Creeper, in 1814 and, as Tree-Creeper, was adopted by the BOU in 1883 to replace Common Creeper as the standard name.

The single-word spelling is now standard and has become the international English name not only for a genus, but for others which illustrate similar habits. The British bird is now known as the Eurasian Treecreeper, *Certhia familiaris*, while the other European form is the Short-toed Treecreeper, *C. brachydactyla*. The one North American species is the American Treecreeper, *C. americana*, which is better known by the old name American Creeper.

The name *Certhia* derives from the Greek *kerthios*, which was a name used by Aristotle to denote a bird of the treecreeper type, while *familiaris* means 'familiar' or 'common'. The form *brachydactyla* simply reiterates 'Short-toed' in a form derived from Greek.

In Australia there are six species which are known as treecreepers, though they belong to an entirely separate family, the Climacteridae. This is a case of convergent evolution, and though the birds are generally larger they have many similar characteristics to the treecreepers of the northern hemisphere. They belong to two genera: *Climacteris*, which means 'ladder', and *Cormobates*, which means 'tree-trunk walker'. I have so far seen four of those: White-throated, Red-browed, Black-tailed and Brown. The other two are White-browed and Rufous. If all bird names were as simple – and boring – as those I should never have started on this journey!

... others climb the wall

The Wallcreeper, *Tichodroma muraria*, is the only member of its family, but is fairly closely related to the treecreepers and nuthatches. It is widespread throughout Europe and Asia, but always at high altitude. Its generic name derives from the Greek *teikhos* (wall) and *dromos* (runner), while the Latin specific *muraria* means 'of walls'. Just how its scarlet wings escaped mention seems totally incomprehensible.

AT LOGGERHEADS OVER THE GREYS: SHRIKES

According to Lockwood, the word 'shrike' first appeared in Turner in 1544. It was then used by Ray and eventually by Pennant, though the latter at first preferred Butcher Bird. Given that Turner received the word second-hand and that it was also a name for the Mistle Thrush (whose shrill notes might well fit the concept of 'shriek'), it was probably a simple confusion. However, the name was standardised by Pennant, who, by noting Great Shrike and Red-back Shrike, confirmed its use as a generic form, which is now used worldwide.

The alternative name, Butcher Bird, was a creation of Merrett in 1667, who translated the Latin form *Lanius*, which is still the generic name. That name was popular and was still often used in books and references which I read as a youngster. It derives from the habit of some species of storing small mammals, birds and insects on thorns. In spite of its popularity, Newton rejected the name and the BOU adopted the current forms in 1883.

On checking whether Coward had listed the alternatives in 1926 (he had not), I noted that Red-backed Shrike was classified as 'summer resident'. It was still resident in my youth, but declined to the point of extinction after 1992. In 2010 a pair became headline news when they bred on Dartmoor. It is still a reliable passage species, however, but rare enough for one to draw a minor local twitch. (In 2009, a juvenile bird on Staines Moor, Surrey, was re-identified as a vagrant Brown Shrike, *Lanius cristatus* – now that *was* a twitch!) The specific name for Red-backed Shrike, *L. collurio*, is taken from *kolluriōn*, a name used by Aristotle, which may or may not have referred to this species.

In recent years it has been difficult to follow the upheaval in the naming of the grey shrikes. Following a widely accepted split, Great Grey Shrike became Northern Grey Shrike, *Lanius excubitor*, and Southern Grey Shrike, *L. meridionalis*. Subsequently Northern Grey was applied only to the American subspecies, for reasons which are explained below, and Great Grey was retained in the BOU and international lists alongside the new southern species. The name Southern Grey Shrike has been replaced in the second edition of the *Collins Bird Guide* by the form Iberian Grey Shrike, which simply refers to the European race of Southern Grey. However, the IOC master list (v6.1) shows potential realignment of the subspecies of the *Lanius excubitor* complex and further likely splits which could separate the North American and Asian forms. Enough to make the head whirl.

The reason for the American adherence to Northern Grey Shrike becomes clearer when its range is considered alongside that of the Loggerhead Shrike, *Lanius ludovicianus*, the southern species. The word Loggerhead suggests an 'over-sized head', which seems to imply that the bird has unusual proportions, though the many which I saw in Florida never struck me as particularly odd in that respect. The specific form *ludovicianus* makes a link with the southern state of Louisiana (based upon Ludovicus for Louis). The use of both Great Grey and Northern Grey is explained by a note on the latest version of the IOC list indicating a need for urgent revision of the complex, with a strong hint of a forthcoming split between the two.

As for the scientific forms, the name *Lanius excubitor* is one of the more useful ones, since *excubitor* means 'sentinel' and reminds us that most wintering Great Grey Shrikes will be found perched on some vantage point. The name *L. meridionalis* tells us that this species will be found further south, which within Europe means mainly on the Iberian Peninsula, as the *Collins Bird Guide* suggests. The American subspecies, the Northern Grey Shrike, is currently *L. e. borealis*. There is no surprise in the name of Lesser Grey Shrike, *L. minor*.

My real favourite among the shrikes is the Woodchat Shrike, *Lanius senator*. According to Lockwood, the English name derives from a word found posthumously in Ray and adopted by Pennant. Yarrell used the form Woodchat Shrike, which eventually prevailed (though with a hyphen) by the time of the BOU's 1923 list. The word Woodchat has no traceable origin before Ray and may have been intended for another more familiar species, given the rarity of that particular shrike in Britain. (My suggestion would be that the Pied Flycatcher was a good candidate, with its chat-like size and structure, and its preference for woodland habitat.) As for the specific name, *senator*, Jobling traces an explanation to Moltoni (1979) that the chestnut-red cap of the bird recalled the red stripe on the togas of Roman senators.

Masked Shrike, *Lanius nubicus*, breeds in southeastern Europe, but winters in northeast Africa, which explains the use of Nubia (an ancient kingdom of that region) in the specific name. Most of the shrikes could be described as 'masked', but in this species the black eye-patch is offset by a white face.

Though Masked Shrike has never been recorded in Britain, other Eastern species are recorded here. The name of the Isabelline Shrike, *Lanius isabellinus*, takes us yet again to the story of Queen Isabella, and the bird is indeed a pale fawn-grey. Brown Shrike, *L. cristatus*, in spite of its name, is not crested. That derives from Edwards' 1747 note of the 'Crested Red or Russet Butcherbird': it is simply that the crown of the bird is russet and stands out somewhat from the plain brown of the mantle. Long-tailed Shrike, *L. schach*, has a curious specific name which derives from the sharp call, rendered otherwise as *scack!* or *tchick!*

We found Woodchat Shrikes to be a fairly common species in Andalucia, but no shrikes have yet to leave a stronger impression on me than a pair of Great Grey Shrikes which were setting up a territory in an Alsace orchard one bright April day, while my son and I took a picnic in the shade. It was such a contrast to those bleak winter encounters with solitary birds on windswept heathland.

WOODCHAT SHRIKE

The names Shrike and Butcherbird have both been recycled in Australia, though the species are not related in any way to the 'true shrikes', and more of those later.

A JAUNDICED VIEW: ORIOLES

The word 'oriole' has its roots in the Latin *aurum* (gold), and its derivative *aureolus*, to arrive at its current spelling via Medieval French. All that makes good sense for a species where the large, showy male is golden yellow with black trimmings that dramatise the effect. An alternative and erroneous explanation was offered by Albertus Magnus (he of the Peregrine Falcon) when he claimed that there was an onomatopoeic link between the name and the fluting song, but that seems just to be a happy coincidence, rather like the song of the Cetti's Warbler.

There is only one species in Europe, the Golden Oriole, *Oriolus oriolus*, which is now the type species for a widespread Old World family. The Golden Oriole is a rare and elusive breeding species in Britain, and my one sighting of a nesting pair at Lakenheath Fen in Suffolk was a true privilege. The family proved easier in Australia, where the Olive-backed Oriole, *O. sagittatus*, and the Green Oriole, *O. flavocinctus*, both crossed my path more than once. The word *sagittatus* (with arrow-like markings) suggests quite rightly that the greenish bird has streaked plumage, but *flavocinctus* (yellow-banded) gives a clue as to why my records recently needed an amendment, since I still had it from my handbooks as Yellow Oriole. That change to the IOC list had removed a clash with another Yellow Oriole which just happened to be on my Trinidad list.

GREEN ORIOLE

One of the recurrent problems with the names of New World birds is that Old World names crossed the Atlantic with the explorers and settlers to become attached to similar species: the word 'oriole' is an excellent example, and one which travelled easily in its English, Spanish and French forms. The New World orioles have many visual similarities to the Old World ones, but they are in fact icterids and of the same family as the grackles, oropendolas and the like, so the second Yellow Oriole is in fact *Icterus nigrogularis* (black-throated). A more northerly species, Baltimore Oriole, *Icterus galbula*, has been recorded as a vagrant to Britain. It has another version of yellow in the *galba* element of its specific name.

Icterus shares the name of the medical condition known as yellow jaundice. From Ancient Greek times the disease and the bird were linked by a folklore which claimed that the sight of an Oriole would cure jaundice, but that the bird would die as a result (see also Icterine Warbler).

A CROWD OF CROWS

Maggie and the pie

Pica pica is stereo Latin for Magpie, but the root of the idea is found in the Latin for 'tar', a reference to the black- on-white plumage of the bird. However, it is the vernacular name which gives the greater interest. When we look a bit closer at the English name, we find that the original form Pie derives from a thirteenth-century French introduction of a name meaning 'pied bird'. That in turn derives from the Latin *pica*. The appearance of the word only coincidentally recalls the name of King Pieris of Macedon, who had nine daughters, the Pierides. He had the temerity to challenge the nine Muses to a contest of skill in the arts. The Muses were daughters of Zeus, no less, so his brazen nerve was punished when his daughters were changed into chattering Magpies.

Shakespeare introduces the term Magot Pie in *Macbeth*, and it has nothing to do with putrid carrion, as one might expect in the context of witches and murder. In fact, the truth is much closer to the story of Jenny the Wren, since the origin is the folksy French creation of *Margot la Pie*. *Margot* is the French version of Margaret (which also contracts to Maggie, Mags, Magot and Mag). A ditty is still extant in French, which begins: *Margot la pie a fait son nid dans la cour à David...* (Maggie the Pie made her nest in David's yard, etc.) The outcome is not so good for Maggie, because David cuts off her legs.

Nowadays, of course, the name has become a generic form, with three *Pica* species now recognised. This makes ours the Eurasian Magpie, while North America has the very similar Black-billed Magpie, *Pica hudsonia*, and the Yellow-billed Magpie, *Pica nuttalli*. A less closely related species is found in Spain and Portugal – and a beauty it is too. Here we need to do yet more reshuffling, since the newly split Iberian Magpie, *Cyanopica cooki* ('Cook's blue magpie') has at last been separated from the east Asian Azure-winged Magpie, *Cyanopica cyanus* (meaning literally the 'blue blue-magpie'). The collector/naturalist concerned was Samuel Cook (1787–1856).

When Magpie was attached to an unrelated Australian species it was for its apparent similarity, as was the use of its name in the much smaller Magpie-Lark, which is neither magpie nor lark. In the first case, the similarity is between the species, but the latter illustrates a phenomenon where 'mapgie', inaccurately, takes on the role of the original word 'pie' (meaning 'pied') for a black and white bird. It is thus applied to such disparate species as the magpie-robins and Magpie Goose.

From little acorns...

The old form *popinjay* is worth a closer look. Having popped up in Chaucer, the word can be found in Shakespeare and later, with its connotations of a gaudily dressed, probably untrustworthy person. It originally meant a parrot and was eventually applied by some to the showy Green Woodpecker. The word 'jay' was an Old French offshoot of the longer word and was introduced to England during the Middle Ages. As a result both words appear in Chaucer's *Parliament of Fowls.* The Jay is indeed a popinjay of a bird, wearing one of the prettiest suits of plumage of any British species, yet with a crow's intelligence and habits. It will cheerfully prey on the eggs and nestlings of other woodland birds, but it gets a worse press than the show-off Great Spotted Woodpecker, which does the same.

But you do have to admire the sneaky little Jay, *Garrulus glandarius.* Garrulous it might be, much noisier than the Waxwing which was once classified with it by Linnaeus. But the second part of its name reminds us that in autumn it pops its acorns (Latin *glans, glandis*) into secret hiding places and recovers only those which it needs. The results can be fascinating: a few years ago a derelict field near my house was awaiting the builders, and, while it waited, Jay-sown acorns sprouted into small oaks all over the field, which was fast becoming a wood. Before building started, the developers brought in a grab-lorry and re-located a hundred or so of these young trees to create a small copse to screen the houses from the main road. With no sense of cause and effect, they then named the entrance road Moorhen Close.

EURASIAN JAY

And it is also the Eurasian Jay, because there are other related species in North America and Europe, such as Blue Jay, Scrub Jay and Siberian Jay. In the case of Blue Jay, the generic name *Cyanocitta* tells us exactly the same as the English name, while *cristata* reminds us that it has a crest. I recall that my first sighting was a serious distraction from viewing the whirlpool below Niagara. I also recall that they were bright blue in the Florida sunshine, but that in the deep shade of bushes they were slate-grey. That is because their colour is caused by refraction of light rather than by pigmentation. Blue Tits have a similar quality.

The Grey Jay, *Perisoreus canadensis*, of North America, and the Siberian Jay, *Perisoreus infaustus*, of Europe share a genus of which the name has obscure origins. We have Bonaparte to thank for creating that in 1831. On face value it seems probable that its root is the Greek *perisōreuō* (to heap up, bury beneath), since these are also hoarders of food. Elliott Coues suggested a link to a Greek bird of augury (via *peri* and *sorix*). Here he had picked on the meaning of the specific name *infaustus* (unlucky) which was derived from an old Scandinavian legend that the bird was an ill omen. Take your pick as to which idea you prefer.

The spelling of Grey Jay aboved posed something of a dilemma. It was tempting to use the American spelling 'gray', since I was looking at my copy of Sibley at the time. However, this is one compromise in which the IOC list favours the British convention. It is a good thing that 'jay' is not similarly variable.

Another related hoarder is Europe's wonderfully named Spotted Nutcracker, *Nucifraga caryocatactes* (I include the word 'Spotted', since its cousin, Clark's Nutcracker, is staring at me from the pages of Sibley). This is a bird which I discuss with mixed feelings: I have never seen one, but am still 'traumatised' by a letter I received from my brother many years ago, in the days when I had no phone. He recounted how, in his northeast Essex garden, he had had a number of birds 'like large starlings'. This was the period when national newspapers were reporting a small vagrant flock of Nutcrackers wandering the area. With no motorway links back then, and a job to do, there was no option of making an impromptu visit to the family – and the birds had moved on anyway. Nowadays he is likely enough to phone me to say that he has a 'funny Starling-sized bird' feeding on the Liquid Amber tree – again! Waxwings seem to find that tree even if there is only a handful in the country.

But returning to that name... There seems to be no problem with Nutcracker, since that is what they do, and *Nucifraga* was originally Turner's translation into Latin of the German *Nussbrecher* (Nutbreaker), so there is little deviation there. But *caryocatactes* is a little more opaque, until we realise that it is a Latinised version of the Greek *karuokataktēs*, which just happens to mean Nutcracker. That surely is a multilingual sledgehammer to crack a walnut.

Jack and the chimney

The *Collins Bird Guide* has two birds perched on a chimney with the caption 'chimney-sweep bird'. These are, of course, Jackdaws, which I recall were sometimes referred to by a similar folk-name, Chimney Daw, when I was young (though I can find no references to that form). Those names come from their use of chimneys as nest-sites: the first name clearly reflects the fact that the young sometimes fall into the hearth below in a flurry of soot. (Incidentally, geese, with their stiff wing feathers, were sometimes dropped down chimneys for the express purpose of clearing the soot.)

The name Daw was first noted as *dawe* in the mid-fifteenth century, but, as an evolved Germanic form, was probably used long before that. The source may well have been onomatopoeic (cf. *caw*). The addition of the word Jack was first noted in 1543, and is assumed by Lockwood to be an onomatopoeic addition which picks up on the 'chack' sound made by the bird. It seems quite likely too that it was also following the pattern of Mag Pie and Jenny Wren, adding a personal name to the species. Jack often simply means small, as in Jack Snipe. Since there was no such thing as a 'large Daw', there is a suggestion of 'jack the lad', something of a juvenile delinquent, since 'jack' also meant 'knave' (as in a pack of cards). The Jackdaw, like the 'thieving Magpie' had an untrustworthy reputation for collecting shiny things. The bird later starred for those reasons in the famous nineteenth-century poem 'The Jackdaw of Rheims', by Richard Harris Barham (alias Thomas Ingoldsby), but the concept is a much older one. The specific name is *monedula*, the Latin for the species: it means 'money-eater' and in turn has its roots in Greek mythology: Arne of Thrace suffered the rather predictable punishment of the gods, when she was turned into a Jackdaw for betraying her nation to Minos in return for money. In short, the Jackdaw's attraction to shiny things seems to have got it into the bad books since time immemorial.

The word Jackdaw entered formal ornithology when it was used by Ray in 1678, and later confirmed by Pennant in 1768. The other formal names of the species are not without the usual complexities. It is Jackdaw in Britain, but Western Jackdaw in some international terminologies, although it sometimes appears as Eurasian or even European Jackdaw. Its congener the Daurian Jackdaw fills the eastern slot.

In like vein, BOU currently retains *Corvus monedula*, though genetic studies have made the case for the two Jackdaws to be placed in the alternative genus *Coloeus*, based on the Greek name for the species, *koloios*. This is preferred by some authorities, the IOC included.

Red in bill and claw

The word Chough seems to be at least as old as Daw, and was noted even earlier, in 1305, as *chogen*. It is likely to be onomatopoeic in origin and seems to have coexisted as a regional word alongside *daw*, with reference to what we now call the Jackdaw. In Merrett's *Pinax* of 1667 he lists: '*Graculus vel Mondula*, a Jackdaw, a Chough', which clearly implies that the two words were synonymous. This is supported by the fact that *graculus* was a word often applied to the Jackdaw, while *mondula* is simply a variant of the specific name eventually used by Linnaeus.

In 1544, Turner noted the term Cornish Chough, and in 1575 there is a record of the use of Cornish Daw, which facts, Lockwood argues, show quite clearly that there was otherwise no separate word for the species. However, later authorities followed Turner, and the name became increasingly detached from the Jackdaw. For once Pennant did not prevail, because his creation of Red-Footed Crow was neatly sidestepped by others. Instead Bewick dropped the word Cornish in 1804, confirming the vernacular name Chough in relation to *Pyrrhocorax pyrrhocorax*, which is still preferred by the BOU.

The generic name *Pyrrhocorax* is a Latin concoction from the Greek *pyrrho-*, which refers to the flame-red colour of the bill and legs, plus *korax* (raven). Chough became a generic English name when discussion concerned a second species, the Alpine Chough, *P. graculus* (in which that old 'jackdaw' word of Merrett's reappears). It seems that Linnaeus left some confusion in the 1766 twelfth edition of the *Systema Naturae* concerning the descriptions of the two species. In their 1912 discussion of this matter, the BOU still distinguished the Cornish Chough from the Alpine Chough. The full international name of the species is now Red-billed Chough, since this is the one clear difference between this and the yellow-billed Alpine Chough. In any case it has a much wider distribution than Cornwall, since it is found widely across Europe and Asia. The Latin *graculus* may be associated with *Gracalus* (from *graucalos*, an ash-coloured bird mentioned by Hesychius).

Grackles and mynas too

The concept carries over into the American name 'grackle', which, as the OED points out, is an Anglicised form of *graculus*, relating to Jackdaws and Starlings. It came into being in 1772 via the form *gracule*, and in 1782 in the current form. The word is omitted by Lockwood, so it may well be an American coinage, though it also sometimes appears with reference to the Hill Myna, *Gracula religiosa*, an Indian member of the starling family, which clearly takes its generic form from the same root. In the Americas, grackles are of the genus *Quiscalus* (once an alternative for *Sturnus*, starling), but these birds belong to the unrelated icterids. As black birds generally approaching the size of a Jackdaw or Chough, they just offered a passing resemblance to the smaller Old World crows, in behaviour as well as in looks. I have seen Common Grackle, *Q. quiscula*, in Florida and Canada; Boat-tailed Grackle, *Q. major*, in Florida; and Carib Grackle, *Q. lugubris*, in Trinidad. The last one gets its name from its 'mourning-black' plumage. Jobling points out that the Carib language contained the word *Quisqueya* (the mother of all lands), and that is suggested as the root of the scientific name.

In the later section on Australian bird names, I look at some issues relating to the White-winged Chough, *Corcorax melanorhamphos*, another unrelated species which exhibits another remarkable example of convergent evolution, similar in size and structure to the European species.

A murder of crows

Dislike is clearly reflected in the collective term 'a murder of crows'. Not only are crows pretty murderous at times, there are plenty of people who would readily murder them for the damage they do to crops and other birdlife. And that leads me to consider where birders sit on the morality fence. To me it is nonsense to apply human morality to wild animals whose lifestyle and opportunities are governed increasingly by man-made distortions of the natural world. I simply cannot understand those who feel that songbirds are 'good' and predators are 'bad'. Predation isn't pretty, but it is natural and it has its own balances. The problem is that we disturb the balances and then blame the birds for the results. When one sharp-witted crow spots an opportunity, others cotton on quickly and a gang-fest may erupt. What we don't always consider is that these are often carrion feeders and that their removal of dead things from the countryside is part of nature's clean-up and recycling process, which benefits us all.

This seems a good point to trot out the old chestnut that 'a Crow in a crowd is a Rook, and a Rook on its own is a Crow'. The only problem is that Carrion Crows have a far more complex social pattern, with some pairing and living in relative isolation, while others remain in the gang. Rooks are in fact more gregarious in general.

ROOK

Both Rook and Crow are Anglo-Saxon, of Germanic origin, as is Raven, and it is probable that they all have a root in onomatopoeia. The generic name *Corvus* is the Latin for Raven, while *corax* says it in Greek. The specific name in *C. corone* is the Greek for Crow, while the *cornix* in the name of the Hooded Crow is Latin for Crow. To give some relief from all this banality we have *C. frugilegus*, which makes the Rook a 'crop-gatherer' (compare *ostralegus* in the Oystercatcher's name). The name is not wholly justified, since the birds eat a greater proportion of invertebrates and earthworms. There is, on the other hand, a reflection of the scientific name in an old country rhyme that I recall from my young days:

> *Four seeds in a hole: one for the Rook and one for the Crow, one to rot and one to grow.*

(In the ultra-wet summer of 2012, the gigantic slugs took number four from my vegetable garden – so much for folklore.)

A heavier sort of legend plays a huge part in the story of the Raven. The bird appears in several Celtic stories, one of which links the Welsh legend of Bran the Blessed with the site of the Tower of London and its tradition of keeping tame Ravens. In Nordic legend, Odin is accompanied by two Ravens, Huginn and Muninn, symbolising thought and memory, so for Viking sailors such mystical qualities were linked to the practice of launching a Raven to test whether land was near. In the Bible Noah too launched a Raven to test whether the flood was subsiding. Again in the Old Testament, Elijah is succoured by Ravens, which then become a symbol of God's providence. Almost every Native American nation had a Raven legend: the Sioux version had a white Raven that spooked the Buffalo, so that the shaman threw the offending bird into the fire and it came out burnt black. The blackest of all, perhaps, is the image of Poe's gothic poem 'The Raven', of 1845, in which the Raven is a symbol of nihilism and death: in context, there is no more chilling line than 'Quoth the Raven: Nevermore!'

It is not too difficult to imagine why the Raven, of all birds, should play such a wide role in so many cultures: it is a large, intelligent bird which has both its attraction as a superb aerial acrobat, and its dark side as a robust carrion feeder. Both admired and feared, its image readily becomes woven into cultures much closer to nature than we are today.

A MURMURATION OF STARLINGS

In the name of the Jay, *Garrulus* underlines the fact that most members of the crow family are well known for being noisy. In fact it is probable that they communicate quite complex information to each other. That is thought to be the case with Starlings too. Yet the widely accepted collective noun for them, 'murmuration', seems only partly appropriate to such noisy birds, while the Latinate form is far from folksy. It first appeared as *murmuracyon* in the fifteenth-century Egerton Manuscript (Palin), though the OED puts it two centuries later in its current form. It was taken up in literary usage in the late nineteenth century to describe the sound of crowds or congregations of people. It seems that it was popularised by the English writer Mary Webb, who wrote in her hugely successful 1924 novel *Precious Bane* that: 'breakfast was ready as soon as our ploughland thawed a bit, and in the stackyard there was a great murmuration of starlings.' In that context the word seems to make sense, because it suits the drowsy conversations of the waking roost, which, scientists believe, is a time when they communicate plans for the exodus of subgroups of the flock. For Mary Webb, a poet, the word had the beauty of alliteration which fitted the awaking day. It has subsequently come to mean the flock or even the gathering of the flock, and is now especially applied to the swirling spectacle of the pre-roost display. At such times, the noise of the flock is the dramatic throb of wings, not voices. 'Mumuration' is a massive understatement of the raw energy of one of these great congregations. I recall one magical evening in the sixties when the awesome sight of possibly millions of swirling birds caused me to follow them towards the copse where they were settling in such numbers that branches broke. The noise was like a waterfall and grew in intensity until it was overwhelming at two hundred metres. I recall the same sense awe while standing on the deck of the *Maid of the Mist* under the spray of Niagara. 'Murmuration' doesn't seem particularly apt for the period of settling down, since the birds have a clear pecking order and quarrel a lot at the tops of their voices. The phrase 'torrent of starlings' would certainly be easier to spell. But thanks, Mary Webb, for a word which proves that birders are poets at heart.

The word Starling itself has a story to tell. The characteristic ending *–ling* is an Old English suffix which indicates smallness (it occurs in Sander*ling* and in Dun*lin*). The Latin word *sturnus* probably gave rise to an old Germanic form, which gave us the word *Stare*, a name by which the bird was known in many parts of the country in the Middle Ages. Originally *–ling* was attached to denote the young bird, and that longer form began to dominate popular usage. Turner used it in 1544, and from then on preference for one form or the other varied from scholar to scholar. Even Latham, who named Kentish Plover and Sandwich Tern, was still using the old form in 1787. It was only in the early nineteenth century that the form we know today became generally accepted. The Latin form *Sturnus* was resurrected for use as the scientific name of the starling genus, and some might argue that *Sturnus vulgaris* is an appropriate name for a vulgar, noisy bird with bad bird-table manners. The original value of *vulgaris* has, however, been distanced by social factors from its simple sense of 'numerous or frequent'. They have become much less numerous now in Britain, so fewer of us get the chance to complain about their table manners.

COMMON STARLING

Spotless Starling, *Sturnus unicolor,* occurs in Spain, North Africa and some Mediterranean Islands. Its name says it all, since it lacks spots and is a plain bird by comparison to its cousin.

Rose-coloured Starling, *Pastor roseus,* is partly rosy-pink only in breeding plumage: in winter it is a dusky grey-pink, even late into the spring, as I discovered when I saw an adult bird at Hordle in the New Forest. The juveniles bear almost no resemblance to adults, but stand out among Common Starlings for their paleness, as another did when we checked a flock in Oxfordshire. The fact that a few of these birds reach Britain each year is due to irruptions of the population. If ever a bird should have a confusion of identity it is this one, since it was referred to as Rose-coloured Pastor in the 1923 BOU list and therefore by Coward in 1926. Its international name is Rosy Starling, while its previous scientific designations have varied from *Pastor sturnus* to *Sturnus roseus. Pastor* means 'shepherd', which may well reflect its place as a bird of rural areas. It winters mainly in India and migrates out into areas of central and eastern Asia. In the 1980s, the province of Xinjiang in China created a large number of nest-boxes to encourage the birds to breed. With grasshoppers forming a major part of the birds' diet, this was a way of controlling locusts: in that way, by the start of the twenty-first century, they had dramatically reduced the need for pesticides, controlled the locusts at the early stage, and reduced pollution.

The Metallic Starling, *Aplonis metallica,* which we saw in the Daintree

Rainforest of Queensland, is another related species. Its iridescent plumage gives it the specific names, though the generic name, *Aplonis* (from the Greek *haploos*, simple, and *ornis*, bird) is a bit disappointing. This was a genus named by Gould as *Aplornis*, but the misspelling was so frequent in use that *Aplonis* is now the accepted form. Fortunately for us, the white belly of juvenile birds showed up well in the dense canopy, otherwise we might have missed them.

A SPARROW FALLS

Sparrow, as might be expected, is a word with ancient roots in the Germanic languages, and it emerged as *sparewe* in Middle English. The word has long had a broad meaning to cover familiar small brown birds of several species, including Hedge Sparrow (Dunnock) and Reed Sparrow (Reed Bunting), and this is still typified by the use of the word in America for birds which are in fact related to the buntings. The case of the Hedge Sparrow was discussed earlier, and the buntings will follow later.

In modern times, the term 'true Sparrow' refers to members of the Old World family, the Passeridae. This includes House Sparrow, *Passer domesticus*, and Tree Sparrow, *P. montanus*. It was Merrett, in 1667, who first listed *Passer domesticus*, House Sparrow. Pennant preferred simply Sparrow for that species, but he was the first to use Tree Sparrow in 1770, as an alternative to the literal translation of Mountain Sparrow, to distinguish a species which had not previously been named separately in English. Yarrell standardised House Sparrow in 1843.

HOUSE SPARROW

The closely related Spanish Sparrow, *Passer hispaniolensis*, made two appearances in England in 2012, close enough to major ports to imply ship-assistance, which is probably the only way such a non-migratory species would arrive. Consequently it remains a rare vagrant.

The jury is still out on Italian Sparrow, which is *Passer italiae* to those who treat it as a full species. For others is it either a stable hybrid of Spanish and House Sparrows or a race of the former.

Rock Sparrow, *Petronia petronia*, belongs to one of several other genera in this family: the scientific name derives from Latin meaning 'of the rocks', but comes via a Bolognese Italian name for the species. The generic name *Montifringilla* is a bit of a surprise within this family, because it means 'mountain finch'. In spite of its names, Snowfinch, *Montifringilla nivalis*, is in fact classified with the sparrows. Its specific name, *nivalis*, reminds us that it is a bird of snowy habitats. There are other species of snowfinch outside Europe, so in a world context this one is the White-winged Snowfinch.

The 'sparrows' of the Americas are the converse of the last case: the term 'New World sparrows' is now used to cover about twenty genera of sparrow-like members of the Emberizidae, which family includes the Old World buntings. As if these birds were not difficult enough anyway.

Sparrows have long been treated as symbols of various sorts. The most obvious use was as something ordinary or down to earth. One cliché, which has died back in its usage over the past decades, is the Cockney Sparrow, the archetypal East End Londoner greeting a pal with such as: 'Wotcher, me old cock sparrer!' The form Sparrer was originally a matter of pronunciation, but the form Spadger, which I certainly recall had wide currency among the children of my generation, was, according to Lockwood, under-written with a tinge of humour. We certainly saw it as slightly rebellious, because our parents disapproved of such slang. A similar phenomenon of slang-names occurred in France, where three such names are still in use. The diminutive singer Edith Piaf, who died in 1963, adopted one of these: her stage surname comes from the Parisian slang equivalent. She was known popularly as *'La Môme Piaf'* (the Little Sparrow). Piaf became well known in Britain for her powerful song *'Rien de Rien'*, which was translated into English as 'No, no regrets!' (I had my own birders version: 'No, no egrets', but the egrets obliged by appearing here by the end of the century.) Nowadays Sparrows have all but disappeared from parts of London, but are still seen as relatively common in Paris.

A second symbolic role served by sparrows was to represent 'smallness', 'insignificance', 'commonplace' and the like. The title of Wilbur Smith's novel *A Sparrow Falls* derives from a Biblical passage in Matthew which refers to the fact that even a sparrow cannot die without God being aware. That, however, did not stop farmers paying bounties for the killing of sparrows, which at one time became major crop pests in this country. Some villages even formed 'sparrow clubs'. Yalden and Albarella state that as many as 100 million were killed in Britain between 1700 and 1930, a period of great agricultural intensification which suited the species. In recent years both House Sparrows and Tree Sparrows have declined dramatically in the countryside, probably because of increased use of agrochemicals since World War 2 and a reduction in waste grain. Sadly their symbolic value here has declined with their numbers. Their introduction into the former colonies led to problems in both North America and Australia, though their populations now seem to be less of a problem, possibly for similar reasons to our own. It is still pretty likely that House Sparrows will be among the first species noted at an entry airport in Florida, Canada or Australia.

CELIBACY, THISTLES AND PINE PARROTS: FINCHES

The origin of the word 'finch' is probably a Germanic rendition of the 'pink' call of the Chaffinch: there are parallels in the Dutch, *Vink*, and the German, *Fink*. However, there are several genera represented within the Finch family, the Fringillidae.

Chaffinch and Brambling

The generic name *Fringilla* is simply the Latin word for finch. That genus includes our Common Chaffinch, *F. coelebs*, and Brambling, *F. montifringilla*.

As for the Chaffinch itself, my earlier discussion of the name of the Lapwing, *Vanellus vanellus*, used the word 'chaff', which is the discarded husk of the ear of grain. Inevitably some grain might be found in the chaff-heap of an old-style farm, so the ground-feeding finch would search the chaff for that food source. (In the 1940s, I lived for a time on a farm which still used pre-war methods of harvesting – horse-drawn binder, sheaves, grain-stack and threshing machine. Our chaff was saved to provide litter for the chicken-house.)

The specific name of the Chaffinch is *Fringilla coelebs*. Now that sort of name is typical of the little secrets which started me down this trail. Linnaeus noted that hen Chaffinches wintered further south than the males (*femina sola versus austrum migrat*) and therefore named the species for its winter 'celibacy'.

Brambling, *Fringilla montifringilla*, revives the generic name of the Snowfinch. That concept also echoes one of its popular names, Mountain Finch. But it is the apparently simple word Brambling that Lockwood pins down in a surprising way. Quite rightly, he challenges the current form of the word, which first appeared in Turner in 1544 as Bramblyng, on the grounds that the species has little to do with brambles. Instead he unearths an older form Brandling (which is in evidence in a note of Pennant's), and which earlier described creatures with brindled markings, one element of which is black (burnt, branded, as in the Black Brant). The element *–ling* is a diminutive, as in Star*ling*. He even goes so far as to suggest that Brandling should be revived, considering Brambling to be an erroneous transcription. Pennant, however, had made the word the standard usage in 1776.

BRAMBLING

The thistle birds

The second genus within this family is *Carduelis*, the meaning of which is found in an alternative name for the (European) Goldfinch, the Thistle Bird, which was used by Ray. It is based on Latin *carduus* (thistle), which is a favourite food of the species, *Carduelis carduelis*.

The Citril Finch, *Carduelis citrinella*, has names which tell of yellowish colour, while its close relative the Corsican Finch, *C. corsica*, clearly relates to the island on which it is endemic.

The Common Linnet, *Linaria cannabina*, has now been given a separate generic name from the Goldfinch. The specific name *cannabina* is a link with common hemp, a plant of the genus *Cannabis*, while Linnet comes via French (*linette*, *c.*1530), from the flax plant, *Linum* (whence also *Linaria*). Both of these plants are grown for their fibres and seeds: in the case of flax, these are linen and linseed. It is of course the bird's liking for the seeds of both plants which leads to the names.

The formal use of the word Common (sometimes Brown) before the word Linnet is more of historical interest than a necessity. It originally distinguished the bird from Mountain Linnet, as the Twite was once known. Hartert *et al.* ensured the dominance of the old and popular name of Twite in 1912. This name was first evidenced in 1562. Mountain Linnet had been created by Ray and was favoured by most ornithologists, including Yarrell. Pennant kept Twite in the frame, and the BOU included both names in 1883, but only Twite appears in Coward in 1926 and thereafter. The scientific name, *Linaria flavirostris*, means 'yellow-billed', which is a relevant feature for the identification of the species in winter. Twite is a version of 'tweet' – and bird names don't come simpler than that!

The name 'siskin' is first seen in Turner in 1544. He took it from a German origin, which Lockwood considers to have its roots in a Slavonic language and to have travelled with the cage-bird trade. The siskins too have recently been moved from *Carduelis* to their own genus, with the British bird as *Spinus spinus*. The Greek word *spinos* was an unidentified bird mentioned by Aristophanes and others. We need Eurasian Siskin to distinguish it from other related species: for example several *Spinus* finches also occur in North America, including the Pine Siskin and three goldfinches. American Goldfinch, *Spinus tristis*, is a step removed from the European one, and has been a regular for me in several parts of Canada. The breeding male is a striking golden-yellow offset with black, a suit which belies its name *tristis*, meaning 'sad'. That word is a bit of a mystery, since the species was originally named *americana* by Catesby, and none of the other (mostly French) names listed by Linnaeus even hint at the concept of *tristis*. His notes show that he might have been influenced by the black head and 'dark' wings (*alis fuscis*), but it is hard to imagine that name arising from a specimen in hand. This is, after all, a bird which would rival the Golden Oriole for show.

The redpolls

These were, until recently, classified as *Carduelis* finches, but they have now been given the genus name *Acanthis*, which is derived from a word used by Aristotle for a sort of finch. Needless to say, there is a transmutation legend behind the name, since Acanthis was a sister of the ill-fated Anthus, he of the pipit legend. On her brother's tragic death the entire family was transmuted into various bird species, and Acanthis became a finch.

The word 'redpoll' simply means 'red-head' (the 'poll' element also appearing in the world of politics as a form of head count for voting or taxation purposes). The name appears in Albin in 1738, and was rivalled by Red-headed Linnet, a form derived from Ray and used by Pennant. The evidence from Coward is, once again, that the modern word was in use at the start of the twentieth century.

Coward does in fact list no fewer than six subspecies, but today there is massive confusion. Some recent research places most, or all, of the genus in the melting pot. At present, there some disagreement about whether Lesser Redpoll is a full species, *Acanthis cabaret*, or merely a subspecies of Common Redpoll, *A. flammea*. It is currently a full species in the BOU list, while the *Collins Bird Guide* still treats it as a subspecies. The current IOC list also has it as a subspecies and suggests that this is a matter still to be resolved. For an inexperienced birder walking into this area of the name minefield, there is the added confusion of the use of the popular name Mealy to describe the 'dipped in flour' appearance of the nominate race of Common Redpoll, which is far from common in Britain.

In principle there should be no problems with Arctic Redpoll, *Acanthis hornemanni*, which is certainly considered to be a separate species. In North America, however, it is known as Hoary Redpoll, and it does have two subspecies – Hornemann's, *A. h. hornemanni*, and Coues', *A. h. exilipes*, both of which may turn up in Britain in winter. At the start of 2012 it was possible to see Lesser, 'Mealy' and Coues' all feeding in one Norfolk garden, and I joined many others to witness the spectacle. The individual differences were absolutely clear to all observers.

As for the scientific names, the Lesser is *cabaret*, which comes from Brisson as the name for 'a sort of finch'. The nominate race, or 'Mealy', is *flammea*, meaning 'flame-coloured', to record the breast colour of breeding males. The trinomen of Coues' Arctic, *exilipes*, means 'slender-footed'. Jens W. Hornemann (1770–1841) was a Danish botanist, and Elliott Coues (1842–1899) was a founder-member of the American Ornithologists' Union. It seems ironic that Coues' Redpolls appearing in Britain will be of Scandinavian origin, while Hornemann's will almost certainly originate from the west.

Greenfinch

The Greenfinch, *Chloris chloris*, is the sole member of its genus in Europe. In a worldwide context it is the European Greenfinch, since there are four other species, all found in Asia. The bird's scientific name could be described as twice as green as its common name. A form of the name appeared as early as 1532 as *grene fynche*, and it was recorded by Ray in 1678 as Green-finch. Ray also recorded Green Linnet as a local name, but Pennant eventually preferred the current name.

The American vireos take their name from a Latin word which possibly originally referred to the Greenfinch.

GREENFINCH

Serins

The Serin, *Serinus serinus*, comes to Britain (infrequently) from across the Channel, as does its French name. The curious thing about this name is that it may have been a French corruption of the Latin *citrinus* (lemon-yellow), which makes its Modern Latin form appear a little contrived. This is strictly the European Serin, to distinguish it from a couple of other species, and it shares its genus with *Serinus canaria*, the Atlantic Canary. Now, in that case, did the bird get its name from the islands, or the islands from the bird? The answer lies in the fact that the word *canis* (dog) gave its name to the *Isla Canaria*, and that island is at the root of the bird's name. But returning to the Serin for a moment, this bird for me is a powerful evocation of the city of Avignon in April, when there was a Serin singing from every available perch in the city. They were common enough too in the cities of Andalucia.

Bullfinch

There are few birds more striking than the male Bullfinch, *Pyrrhula pyrrhula*, though this shy bird tends to show its white rump more often than its gorgeous rosy breast. The name *Pyrrhula* means 'flame-coloured' and comes from the Greek *pyrrhoulas*, a name used by Aristotle for a 'worm-eating bird'. In truth that sounds just a bit more like the Robin. But the Bullfinch is the antithesis of the Nightingale, since his fluty whistle is definitely inferior to his 'whistle and flute'. As to the name Bullfinch, Lockwood traces it to an earlier and shorter form Bull, pointing out on the way that the Bull which tolls the bell in the traditional 'Cock-Robin' rhyme was more logically the bird. He also points out that the bird has a parallel French name, *Boeuf* (*ox*), which suggests a Medieval French input into the English version. There is also the earlier-mentioned fact that some small birds were known as *oeil de boeuf* (bull's-eye, ox-eye), but that seems less likely for this bulky finch. In the simplest terms, it seems obvious to an observer that the bull-necked shape of the bird is a good enough explanation. Turner, in 1544, is the earliest evidence of the combination of 'Bull' with 'finch', and that usage travelled the familiar route via Ray and Pennant into a standardised form.

Our bird is the Eurasian Bullfinch, since there are half a dozen other species in the genus, including the endangered species of the Azores, a much plainer form, whose name, *Pyrrhula murina*, describes its 'mouse-coloured' plumage.

Trumpeter Finch

The Trumpeter Finch, *Bucanetes githagineus*, has been recorded in Britain only very rarely, but it may well be encountered in southern Spain, though I missed it there. The distinctive 'toy trumpet' buzz earns it both the vernacular name and, from Greek, the generic form *Bucanetes*. The specific name *githagineus* relates the bird to the plant genus *Githago* (e.g. Corncockle). The IOC list tells us that the Spanish race is *B. g. zedlitzi*, which records Otto Graf von Zedlitz, a German ornithologist (1873–1927).

Grosbeak

Neither have I yet seen the Pine Grosbeak, *Pinicola enucleator*, but the name is too good to pass by. 'Grosbeak' is simply an Anglicised form of the French *Grosbec* (large-bill), which was adopted by ornithologists from its use by Belon (1555). As for Pine Grosbeak's scientific name, *Pinicola* is 'pine-dweller', and *enucleator* means that it 'removes the kernel from the cone'.

Hawfinch

Formerly known as the Common Grosbeak, the Hawfinch, *Coccothraustes coccothraustes*, has a unique and annoyingly repetitive scientific name, yet Coward, a stickler for trinomials, used it in triplicate. Its Greek root is in *kokkothraustēs*. The kernel of that nut lies in *kokkos* (seed) and *thrauō* (break). The original name comes from Hesychius, but is otherwise unidentified. Gesner first used the Latinised version in 1555. The Hawfinch is the unique member of its genus, but there are six subspecies, which are found as far away as Japan and North Africa.

The current English name was a relative latecomer, since Common Grosbeak was preferred by Pennant and earlier writers. Selby, in 1825, and Gould, in 1837, were followed by Yarrell in 1843 in using a name which refers to the haw, the berry of the hawthorn, and a major food item. In the end it does seem a rather tame name for a bird which has evolved jaws and a bill capable of cracking wild-cherry stones. And this is another bird which gets by on its striking looks rather than its feeble voice. I recently watched one on a song-perch high on a fir-tree: its voice was barely audible, but it stood out wonderfully against the sky.

Crossbill

Common Crossbill, *Loxia curvirostra*, bears an English name which is glaringly obvious in its meaning, but seems to have no traditional root in this country. The nearest this bird had to a folk-name was the form Shell Apple, which Lockwood links to Sheld Apple, an old name for the Chaffinch and which derives from a word meaning variegated (see Shelduck). Crossbill, and its equivalent Crossbeak, seem to be 'book-names' based on Gesner's 1555 creation, *Loxia* (from the Greek *loxós*, crossed), which in turn derives from his native Swiss-German *Krűtzvogel. Curvirostra* means 'curved bill'.

In *Loxia pytyopsittacus* (Parrot Crossbill) the two elements of the specific name mean 'pine parrot'. *L. scotica* (Scottish Crossbill) is self-evident. In the name of the Two-barred Crossbill, *L. leucoptera*, the specific name means white-winged and refers to the wing-bars which provide the English name. This last species is known in America as the White-winged Crossbill, and that is the nominate race. The Eurasian one is *L. l. bifasciata*, meaning 'two-barred'.

Rosefinch

The Common Rosefinch, *Carpodacus erythrinus*, has always eluded me, though I have seen two of its congeners in Canada. Rosefinch is self-evident for the red male. The generic name *Carpodacus* means 'fruit-biter', while *erythrinus* is from the Greek *eruthros* (red).

The common name of its relative, the North American Purple Finch, *Haemorhous pupureus*, is also quite self-evident. However, *Haemorhous* derives from the concept of a 'blood-red rump'. I was introduced to that species in a Quebec garden. The generic name is shared by House Finch, *H. mexicanus*, which has a curious story to tell, since this bird originated from Mexico and the southwest of the US, but was sold in the 1940s in large numbers as a cage-bird (or house-bird) under the enticing name of Hollywood Finch. When a conservation act threatened prosecutions, many caged birds were released in the eastern states, where the birds rapidly naturalised to become widespread throughout the USA and Southern Canada.

HAMMERS AND SPURS: BUNTINGS

The family name, Emberizidae, and the common generic name *Emberiza*, both stem from a German word *Embritz* (bunting). The Reed Bunting, *E. schoeniclus*, shows a further Germanic influence, since the structure of *schoeniclus* is down to Gesner's rendering of a Greek name, *skhoiniklos*, a small waterside bird mentioned by Aristotle, which may well be a good fit for the Reed Bunting.

REED BUNTING

Corn Bunting

It has to be admitted that the beauty of vernacular names, like that of Dunter for the Eider, is that they often reflect the imagination of the locals. While that name reflects an observation, others can reflect emotions, such as fondness, dislike, humour, or even fear. The Corn Bunting, *Emberiza calandra*, has something of a 'cuddly granny' story... The old nursery rhyme 'Bye, Baby Bunting' holds a clue as to what medieval people thought of the familiar little round birds which sang in the corn: they were chubby and harmless like the chubby child (the 'bunting' of the song), so today we have a whole family of birds named fondly after the chubby child. One of its other folk-names was the Corn Dumpling, which reflects the same idea, rather than one of food.

However, the converse of that rural idyll is found in the bird's recently abandoned scientific name, *Miliaria calandra*. We read of ancient royalty feasting on larks' tongues and we still hear that certain Mediterranean cultures think little of skewering song birds to roast as a delicacy. From time immemorial, larks were caught as food items all over Europe: being of a good size, Skylarks (and Calandra Larks) would be sought after. Corn Buntings were considered to be the next best thing, a sort of 'poor man's lark'. To lure them into traps, millet seed was set down, hence *Miliaria* for millet and *calandra* for lark – and no sentimentality this time. Sadly that story is now partly hidden, since the bird is now listed by IOC and BOU as *Emeberiza calandra*.

Yellowhammer

Another familiar bunting, the Yellowhammer, *Emberiza citrinella*, derives its specific name from the diminutive of *citrinus* (citrine). This species holds a very special place in my heart mainly because my own country-born grandmother taught me that it sang 'a little bit of bread and no cheese' (which is still there in Collins). The development of the name Yellowhammer is curious, since it underlines the power of popular forms in the way that Lapwing does. The true origin of the 'hammer' element has its roots in an Old English word of Germanic origin, *ammer* (meaning 'bunting'), which is related to *embritz*. There was some understandable confusion along the way with *amber*, since the rich yellow-brown colour of the fossil resin fits well with the male Yellowhammer's rich rump-colour: the form Yelamber and variants existed in the sixteenth and seventeenth centuries. Other dialect variants included Yellow Ammer, Yellow Ham and Yellow Am-bird. The ornithologists steered the name towards its modern form: in 1678 Ray used Yellow Hammer, but in 1776 Pennant adopted the logical form Yellow Bunting, which Yarrell also used in 1843. That form had the advantage of placing the bird firmly alongside such as Corn Bunting and Reed Bunting, as well as reflecting the original sense of *ammer*. Surprisingly, in 1883, the newly formed BOU preferred the two-word form Yellow Hammer, presumably to retain an element of the folk-name. The founder, Alfred Newton, did have something of a penchant for tradition. It was eventually to revert to the single-word form, Yellowhammer, in the early twentieth century.

YELLOWHAMMER

Cirl Bunting, Rock Bunting and others

Two other close relatives of the Yellowhammer, Cirl Bunting, *Emberiza cirlus*, and Rock Bunting, *E. cia*, both take their specific names from an Italian verb, *zirlare* (to chirp). *Cirlo* and *Cia* are local Italian names. I saw them both in Spain.

The Ortolan Bunting, *Emberiza hortulana*, is rare in Britain but widespread in continental Europe, and is possibly known more among non-birders for its culinary fame, a now illegal practice which both damages the species and does nothing for the positive image of humanity. Both the French-rooted common name and the specific name derive from the Latin *hortus* (garden), via the Italian *hortulane*. The closely related and similar Cretzschmar's Bunting, *E. caesia*, is found in Greece and the Middle East. Its specific name means 'bluish-grey', which distinguishes the head and neck colour from the greenish-grey of the Ortolan. The German naturalist, Phillipp Cretzschmar (1786–1845) is commemorated in the common name.

A number of other species stray into Britain during migration. In general there is no mystery in their names. Black-headed Bunting, *Emberiza melanocephala*, means the same in English and Latin, as do Little Bunting, *E. pusilla*, and Rustic Bunting, *E. rustica*. Yellow-breasted Bunting, *E. aureola*, is literally golden in its Latin form. Pallas's Reed Bunting, *E. pallasi*, takes both its names from the German naturalist Peter Pallas, whose name has already occurred a number of times.

And those American sparrows?

American representatives of the Emberizidae include some twenty genera and forty-eight species, including the New World sparrows, juncos and towhees, a few of which appear in Britain as vagrants. The most famous of these on recent years was the White-crowned Sparrow, *Zonotrichia leucophrys*, which graced Cley in Norfolk in early 2008. Grateful birders (of which I was one, since the club was birding in the area) threw so many coins and notes into the collection bucket that the local church gained a new stained-glass window, in which the bird is depicted. The bird's scientific names paint its portrait by speaking of 'banded hair and white eyebrows'. My only other British sighting of an American sparrow was a Dark-eyed Junco, *Junco hyemalis,* in the New Forest – and I have to say that these two are relatively easy species for British birders, as is the White-throated Sparrow, *Zonotrichia albicollis.* Suffice it to say that I have struggled many times with some of the less well-marked species on the other side of the Atlantic. All three of those mentioned are highly migratory within America and are therefore vulnerable to seasonal storms. In fact, the specific name of the Junco, *hyemalis,* derives from the fact that it arrives in the States in winter, where it is also known as the Snowbird – as are the Canadians who winter in Florida.

Longspurs

In 2008, Alstrom *et al.* proposed that two bunting genera should be removed from the Emberizidae and into their own family, the Calcariidae. In winter we in Britain might be lucky to encounter two species, which have names that relate to their 'spur' (Latin, *calcar*), the rear-facing toe. The first is the Lapland Bunting, *Calcarius lapponicus,* which is known internationally as the Lapland Longspur, a name which explains why *calcar* is important. *Lapponia* is of course Lapland. In the name of the Snow Bunting, *Plectrophenax nivalis,* the spur is the Greek, *plēktron,* while *phenax* implies 'an imposter'- or to put it another way, it does a pretty good imitation of a the spur of a cockerel. The specific name *nivalis* means 'of the snow'. This separates it from its one close relative, *P. hyperboreus* (northern), which is known in America as McKay's Bunting.

SNOW BUNTINGS

(Before I leave the buntings and this phase of the story, I am reminded that typographical evolution has produced the Red Bunting. It was just too good a bird to waste. Its habitat is, of course, the fringes of firing ranges. It has a distinctive jizz, since it flutters only in a stiff wind. And, if while looking for one, you see a Missile Thrush, make sure you duck!)

NEW HORIZONS

THE BLUE MOUNTAINS FROM KATOOMBA

In this last part I look at further issues related to bird names in North America, Australia and Trinidad, all of which have intrigued me in one way or another.

One obvious theme is that of native forms which have been adopted, adapted or absorbed into the formation of English or scientific names. It is in fact particularly difficult to find traces of these in North America, though they are somewhat more frequent in the formal Australian names. One reason for the difficulty is the lack of a written language in both cases, but the scarcity also relates to the fact that bird names, like history, were written by the dominant culture, which either suppressed or ignored native cultures. North America has had longer than Australia for the evidence to fade, but that factor does not seem to apply to the vocabulary of South American species, which seems to be far richer in native words, with many examples showing distinct local influences.

An offshoot of this first element is that of 'local colour', which was a term used by the Romantic writers of the early nineteenth century to convey the atmosphere and texture of exotic environments. It was a way of painting word images which evoked the details and differences of overseas (or historical) locations. In that respect I enjoy seeking out the names which reflect such elements. Those are not always 'native names', but may equally be colonial names invented or used long before the ornithologists took hold of matters: after all, the settlement of North America began in earnest in the early seventeenth century, long before Ray and Pennant began to steer formal English names towards their current form. As a result many of the differences already discussed in the vocabulary of such as raptors and seabirds are down to a long separation of the cultures of Britain and North America. That phenomenon seems not so pronounced in Australia, which has had less time as a separate Anglophone entity, and a relatively short period before the collection, naming and cataloguing of their wildlife began in earnest. However, the wildlife on the other side of the Wallace Line* is hugely different, and what struck me when I first went there was the fact that I was seeing whole new families of birds: that is colour enough!

The problem is, however, that the drive for standardisation and for universal handles means that some colour fades. Macdonald points out that 96% of Australian species had been named and listed by 1910, but that a series of wholesale reviews had been undertaken by the time he was writing in 1987, with the result that over 70% of the original vernacular names had been formally changed. When we consider the influence of such as Pennant on the shaping of names, we realise that prejudice against 'folksy' forms can erase many colourful names and increasingly replace them with rational 'book-names' (to use Lockwood's term). North America started down that trail rather earlier than Australia, with a great deal of work done in the nineteenth century. Later, the squeeze to internationalise names, which I discussed earlier, has not helped to preserve local colour. The third theme is, therefore, a comment on some of those names created and agreed by the ornithologists. Some are straightforward, helpful, even; some are confusing; and some are plain irritating.

In the three chapters which follow, these themes are interwoven, as I consider a few more things which have interested me in those three areas of the English-speaking world where I have been to watch birds.

As for structure, this section is more thematic than schematic, so I don't intend to lose the trails through this particular wilderness.

CROSSING THE GREAT DIVIDE: NORTH AMERICAN NAMES

I have long been intrigued by the way in which a shared language links us to North America, yet just how far we can be separated within that language. That is part of the paradox in which common roots of a cultural heritage link us to the Americans and Canadians on the one hand, while on the other the evolution of those cultures has led to distinct shades of difference. In some ways there are parallels with Australia, but in the latter case the bedrock of the separation is some two centuries younger. In both cases, the English Language travelled with the explorers and settlers. What we must remember is that the language used by the

* Alfred Russell Wallace (1823–1913) was a British zoologist who first described the zonal division which separated the fauna of Asia from that of Australasia. The Wallace Line passes through the islands between the two continents. Wallace's quite independent development of a theory of evolution prompted Darwin to publish his own work, which he had long suppressed for fear of upsetting a dominantly Christian-fundamentalist British establishment and, not least, members of his own close family.

earliest arrivals in America was not the English we speak, read and hear today. Christopher Newport landed the founders of Jamestown in 1607 and Christopher Jones sailed with the ship *Mayflower* and the Pilgrim Fathers in 1620. These people spoke a more down-to-earth version of the language of Shakespeare and of the King James Bible, so it was inevitable that, over the following centuries, American English would evolve in subtly different ways, developing its own regional dialects and vocabularies. The American War of Independence underpinned this process with an element of national pride which took firm ownership of its intellectual properties. From these elements are born the differences of usage and nomenclature which underline the ways in which our common language can sometimes divide us.

By coincidence both Newport and Jones had links to my home town of Harwich, so their stories were part of my local culture. Almost four hundred years later I made it across the Atlantic to discover for myself that all was not so simple in the naming of birds.

Same difference

One thread of 'sameness' has already been discussed. By that, I mean those names like 'loon' and 'murre' which were adopted in America from forms of English vocabulary which then became obsolete here and are therefore unfamiliar when we meet them in North America. The second 'sameness' is also a paradox, since this is the inevitable case where familiar English words became attached to birds which are similar, but unrelated, to those for which they were created. Names such as 'vulture', 'robin', 'oriole', 'redstart', 'blackbird' and 'warbler' are of this sort: when used in North America they refer to something quite different. While Australian orioles belong to the same family as the Eurasian Golden Oriole, the North American 'orioles' belong to a wholly different family, the Icteridae, as do the 'blackbirds'. North American 'redstarts' are in fact 'wood-warblers', a general term used to distinguish the New World species from those of the Old World. In short, there are a good many aspects of bird names which require some adjustment when we start to bird across the pond, and many of those have been discussed already.

Fading footprints

First, however, I wanted to track down some evidence of the influence of Native American names, and in truth I found that four hundred years had taken their toll. I did earlier discover one example in the word 'dowitcher', and I have unearthed a second example in the name of the Sora (a species of rail). Even for that name, however, the evidence is weak, with most dictionaries claiming no more than Choate: that it is probably a name of native origin. After those two examples the trail would run cold fairly quickly, were it not for the names influenced by matters south of the Mexican border. With few exceptions such species all have an extreme southerly range in the United States, though the Turkey Vulture, with a link in the scientific name, is an exception.

The best-known of such names belongs to the curious relative of the cormorants, the Anhinga, *Anhinga anhinga*. The name passed into Spanish and Portuguese from the Tupi people of modern Brazil. It is also often called a

Snake-bird, for the fact that it swims with just its sinuous neck projecting from the water. Its African and Australian relatives are also known as 'darters', from the spearing thrust with which the birds take fish. One popular name cited by Choate is the Water Turkey, which is a passable approximation for a large dark bird which rests by water.

Two species of Ani, relatives of the cuckoos, also take their name from a Tupi word, *anim* (sociable). We missed Smooth-billed Ani, *Crotophaga ani*, in Florida, but eventually caught up with the curious, heavy-billed black birds in Trinidad. *Crotophaga* is an interesting idea, since it means 'tick-eating'. Yummy!

The Northern Crested Caracara, a raptor, has yet another Tupi name, which reflects its raucous voice and is said to mean 'noise-mangler'. The Venezuelan Amerindians get in on the act too in the specific name, *Caracara cheriway*.

A second raptor to bear a name of similar origin is the Californian Condor, *Gymnogyps californianus*, at least in the 'condor' element, which originates in the Peruvian Quechua language as *kuntur*. The generic name, meaning 'naked vulture' is a reference to the head and neck rather than to the bird as a whole.

There is less subtle evidence in the name of the Montezuma's Quail, *Cyrtonix montezumae*, which recalls the hapless sixteenth-century Aztec Emperor Moctezuma II, who was conquered by Cortes. The Peruvian Incas account for Inca Dove, *Columbina inca*, and Inca Jay, *Cyanocorax yncas* (Green Jay in many field guides – which just happens to be a 'blue raven' in its generic form). The Ruddy Ground-Dove is *Columbina talpacoti*, which specific name may or may not be a word of Tupi origin.

LIMPKIN

In the case of Limpkin, *Aramus guarauna*, another Tupi word provides the specific name *guarauna*, originally meaning 'a black marsh bird'. This is a species found widely in South America, but Florida's bird is the subspecies *pictus* (painted), which describes the white splotches on the dark plumage. Having been given the impression that it was a secretive bird, we were surprised to encounter one in the open in a Clearwater park, not once, but two years in a row. It is a wonderful crake-like bird with aspirations to become an ibis. The odd English name, Limpkin, is more straightforward than one might expect, because the bird's odd gait promoted the idea, while the diminutive suffix *–kin* seems to imply an element of affection or maybe pity. The attachment of the generic word *Aramus*, once used by Hesychius for a 'heron', is misleading, since this species is quite unique and unrelated to the herons.

The French connection

French also comes into the picture in the names of two birds.

The first, becard, is the name of a widespread South and Central American family, which is represented in border areas by the Rose-throated Becard, *Pachyramphus aglaiae*. Becard derives from the French word *bec* (beak), and is applied as *bécard* in modern French to a salmon with a hooked upper jaw. A proportionately large bill accounts for the name in the birds. The generic name *Pachyramphus* also means thick-billed, while the specific name is from Aglaïa, one of the Three Graces of Classical mythology.

The name of the Verdin, a small American relative of Penduline Tit, is a really odd one. Choate only gets half of the story, since he quotes Alexander Wetmore to assert that 'it is the French for yellow head'. The truth is that the bird on which the name is modelled does have a yellow head, since the French lexicographer, Littré, records *verdin* as a synonym of *bruant* (Yellowhammer). That species has an olive-brown back, which just about accounts for an obsolete name based on *vert* (green). In other contexts the French *verdin is* applied to several green species of Asia, while the similar Spanish form *verdón* is also applied to a number of green-backed birds. It is certainly a misleading name to choose for the dominantly grey and yellow American bird. Fortunately the scientific name puts matters right very firmly, since the American Verdin is *Auriparus flaviceps*, which means literally the 'yellow-headed golden-tit'.

Spanish comes into the story too in the name of a falcon which is sometimes noted in the extreme south. It is the Aplomado Falcon, for its 'leaden' blue-grey plumage. The specific name, *Falco femoralis*, reminds us that its upper leg is well feathered.

Echoes

There may be traces of Native American words in some echoic names, since the roots of such names could belong to any culture. In any case, the settlers who created many of these may well have been from almost any non-British European stock. America was, after all, the melting pot of Europe, as exemplified in E. Annie Proulx's symbolic work *Accordion Crimes*. Forms such as Chachalaca, Towhee, Pewee, Whip-poor-will and Poorwill are obviously imitative. Perhaps the most interesting name of this sort is that of the Saw-whet Owl, since the call reminded people of the screeching sound of the file used to sharpen (or whet) a saw. It is easy enough to imagine the creation of such names, since words like Cuckoo and Peewit were a precedent: for that reason alone, it is no great surprise that a two-note call resulted in the same name, Pewee, in both America and Australia, though for wholly different species.

However it is almost a disappointment to find that seemingly more interesting forms like, Bobolink, Bobwhite, Dickcissel, Phoebe, and even Chuck-will's Widow belong in the same category. There is little surprise, though, that familiar names like Chuck, Will, Bob, Dick and even Phoebe get involved, since that again echoes the tradition which gave us Robin Redbreast and Jenny Wren. Among those names it is worth noting an older extension of Bobolink to Bob O'Lincoln, which rather underlines that pattern.

With all of these names, however, the input of ordinary settlers and farmers remains very evident, whether or not they Anglicised any native echoic forms. As we shall see later, Australian scientists bulldozed aside many such names during the first half of the twentieth century, replacing them with 'book-names'. This clearly did not happen quite so widely in North America, as this cluster of names clearly shows. However, the scientists have left some pretty heavy footprints.

In nomine...

The influence of Latin is quite marked in some of the extant vernacular forms, which clearly indicates that the scientists were highly influential in that area.

Bonaparte was responsible for one curiosity. When faced by the need to name a new species of bird from the Mexican border country he was reminded of the European Bullfinch (*Pyrrhula*) and Crossbill (*Loxia*). The result is the concoction of the name Pyrrhuloxia, which has survived as a most unusual vernacular form. Its scientific name is *Cardinalis sinuatus*, where the word *sinuatus* describes the strikingly rounded bill. It shares with its more familiar congener, Northern Cardinal, *Cardinalis cardinalis*, a name which compares the bright crimson plumage of the male to the robe of the Roman Catholic dignitary. Pyrrhuloxia, however, is a more subdued grey bird trimmed with red.

CARDINAL

If that last name hardly trips off the tongue, there is a further species, also of the extreme southwestern states, which bears the vernacular name Phainopepla, which in this case is also the generic name for a small number of species known as silky-flycatchers. The name in fact means 'shining cloak', which is reinforced by the specific name *nitens* (shining). Presumably this is a case where the bird is known to so few in North America that it has never justified an Anglicised name.

Vireo is a name of similar origin, though one which sits much more comfortably on the tongue and is also used widely as a familiar name. In Aristotle this was possibly a Greenfinch, and it was used by Turner as such in 1544. Its adoption for the American genus is down to Vieillot in 1807. Red-eyed Vireo, *Vireo olivaceus* (greenish), is a vagrant to Britain. (A Canadian bird once descended from dense canopy in response to my first attempt at 'pishing', and I have never been so amazed in my life.)

The name kingbird is conceptually linked to the generic name *Tyrannus* and to the term tyrant-flycatcher. Eastern Kingbird, which I found once or twice in Florida, is *Tyrannus tyrannus*. Members of this family exhibit extremely aggressive behaviour to protect their nest sites, literally 'ruling the roost' in a tyrannical way. A mostly hidden, crown-like crest reinforces this metaphor and recalls that of the tyrant Hoopoe which got the Lapwing into trouble.

The Flammulated Owl, *Psiloscops flammeolus*, has a cumbersome Latinate vernacular name which describes the flame-like markings of bright rufous in its otherwise monochrome plumage. Ferruginous Pygmy Owl uses a more familiar Latinism for 'rusty-brown', which also occurs in the name of the European duck – but more of that owl later.

And while we are concerned with the influence of the Classical languages on American names, this is a good moment to consider the term icterid, which I used earlier when discussing the evolution of the name grackle. The Icteridae are members of a family which takes its name from the Greek *ikteros* (jaundice-yellow). Included within the family are the American orioles and meadowlarks, which do sport plenty of yellow, but they also include grackles, American blackbirds, Bobolink and cowbirds, which tend to be predominantly black. Further south there are the oropendolas and caciques, which I will revisit later.

Plain English

In many ways plain names reflect the work of the scientists almost as much as do the scientific names. They could be spectacularly inventive at times, as I discuss later in the names of the hummingbirds. However, the 'book-names' which so often occur are sometimes of interest because they represent something which is unfamiliar to the British birder – that touch of the exotic, perhaps – so this section touches on some of those.

The skimmers, relatives of the gulls and terns, feed by skimming the surface of the water with their grotesquely developed lower mandible. An eighteenth-century alternative, cutwater, which was applied to the American species, offered a rather more evocative description of the feeding action. We occasionally saw Black Skimmer, *Rynchops niger*, feeding in Florida, but they were more often seen roosting, both there and in Trinidad, sometimes with the massive bill resting flat on the sand. The generic name describes this odd-shaped bill in a curiously inverted way: rather than speak of the elongation of the lower mandible, it suggests that the upper mandible is cut off – from the Greek *rhynchos* (bill) and *koptō* (cut).

As I discussed earlier, some of the American woodpeckers have interesting names and stories, but there are others which are worth a mention here. Sapsucker is a collective name for four species of woodpecker which drill holes and eat the sap and insects which accumulate around the wound. The generic name offers a Greek version of *martius* in *Sphyrapicus* (hammer-pecker). The one eastern species, Yellow-bellied, *S. varius*, has always eluded me (it clearly lacks the courage to show itself!). However, Downy and Hairy Woodpeckers have become familiar. The names of those two look odd at first sight, but relate to tufts of fine feathers above the bill. They share the generic name *Picoides* with the (Eurasian) Three-toed Woodpecker which was discussed earlier, but their respective specifics *pubescens* and *villosus* mean exactly the same as their English names.

Acorn Woodpecker takes its name from storing acorns in neatly chiselled slots in tree-trunks. What is more, it garners these stores in a complex social arrangement, which is unlike that of any other woodpecker species. Its scientific name, *Melanerpes formicivorus*, makes no mention of these odd habits, since it simply means that it is an 'ant-eating black creeper', which is not inaccurate for a species which feeds its young on insects and also catches some in flycatcher style.

Among the icterids mentioned earlier are the cowbirds, which, as might be expected, took their name from feeding near cattle. The present name is in fact a contraction of Catesby's original creation of Cowpen-bird, because he first noted them in cattle pens. An old alternative name was Buffalo Bird, since the species was known to follow the herds of American Bison in the Great Plains. As with Cattle Egrets, nightjars and Yellow Wagtails, the attraction is, of course, the insects disturbed by the grazing animals. Indeed, Elphick, Dunning and Sibley report that some biologists have suggested that the bird's brood-parasitism evolved to allow the adults to leave others to raise their young while they followed the itinerant herds. However, they also point out that, with ancestry in South America, it is more likely that the breeding predisposition of the birds made them more suited to follow the Bison. Brown-headed Cowbird, *Molothrus ater*, is a familiar species which we have seen regularly in Florida in winter and in Canada in the summer. The word *Molothrus* is an odd one, since it speaks of a 'struggle to impregnate'. Choate states bluntly that this was a mistake, since it should have read *Molobrus* (a greedy person or a parasite). Coues stated that Swainson wrote it wrongly so often that it became the accepted form. Brown-headed Cowbird lives up to its simple vernacular description: it is a quite handsome bird, with glossy blue-black body and chestnut head.

Another family, the Mimidae (mimics) includes the Mockingbird. This is a simple yet effective name, which has a special magic for me. I grew up with the songs of American bass, Paul Robeson, and my favourite was one which opened: 'Lindy, did you hear the Mockingbird sing last night?' I actually got to hear Robson sing that number live in the late 1950s, but it was almost another forty years before I heard the bird sing in the Florida night. Its 'mocking' is imitative rather than cynical, and the variety and sweetness of the song rivals the Nightingale or Blackbird. Its scientific name is *Mimus polyglottos*, which echoes in part the name of Europe's Melodious Warbler. We saw the Tropical Mockingbird, *M. gilvus* (pale yellow), in Trinidad, though I found it less attractive than its alternative name, Graceful Mockingbird, suggests. Perhaps it was because I had long been very much more attached to the Northern Mockingbird, as the Florida bird should be called.

The Grey Catbird, *Dumetella carolinensis*, sings well too, but also creates sounds like the mewing of a cat. These are skulking little birds, coloured like the shadows, and bearing a generic name, *Dumetella*, which links them to the thickets. I have seen them in Florida palmetto scrub in winter and shrubs in the

Montreal Olympic Park in summer. The Australians also have a catbird, but that one is related to the bowerbirds.

The gnatcatchers apparently catch gnats, but that seems also a good way to separate this different family from flycatchers. The Blue-grey Gnatcatcher, *Polioptila caerulea*, became familiar in woodland in Florida, where its behaviour was part tit, part warbler, part flycatcher, but distinctive with its pale grey shades. Its scientific name repeats the concept of 'blue-grey plumage'.

American wood-warblers

In Florida, I once came across a pair of drab warblers with striped heads which were vaguely reminiscent of the European Sedge Warbler. They were foraging in the branches of some low scrub and turned out to be Worm-eating Warbler, *Helmitheros vermivorum*. The scientific name is a concoction of the Greek – *helmins* (worm) and *–thēras* (hunting) – and Latin – *vermis* (worm) plus *–vorus* (eating). So one would well imagine that this bird might spend all its time on the ground grubbing for worms. Wrong! They may forage in leaf litter, but they hardly ever eat earthworms: rather, they probe clusters of hanging dry leaves for insects and caterpillars. That fact reminds us that 'worm' has a wider meaning, as in 'inch-worm', a type of caterpillar, but it does seem to be an over-laboured and somewhat misleading name. At least the Spotted Pardalote really is spotted.

The Northern Parula is better named, because *Setophaga americana* derives from the Greek sēs (moth) and *–phagos* (eating) and the bird does at least eat insects, including moths. Most American wood-warblers belong to that genus. Spring passage in Florida is very special, and it brought me my first close-up views of this pretty bird. The origin of Parula is simple, since it derives from *parus* (tit) and so was 'tit-like'. I fell for this beauty the moment I saw it. It was one of those sunny Florida days during spring migration, and I was seated at a picnic table at Fort De Soto alongside a tree full of passage migrants. I thus saw my first Parula well enough to appreciate all details of its subtle pastel-box of plumage.

While down-to-earth names can work very well, I find that the names of a few of the American wood-warblers are profoundly irritating for being so pedantically descriptive. My pet 'hates' are the Black-throated Blue Warbler and its ally the Black-throated Green Warbler: after all, their scientific names, *Setophaga caerulescens* and *S. virens* offer the essential colours so neatly. The name Black-and-white Warbler seems both clumsy and banal. Even the scientific name, *Mniotilta varia*, does that bird an injustice: Vieillot wanted the Greek-based generic name to mean 'moss-plucker', but he somehow plucked the wrong word out off the dictionary to get 'seaweed'.

Simpler solutions do exist: Common Yellowthroat is, after all, in keeping with Lesser Whitethroat, with just enough description, and you couldn't do a lot better than Yellow Warbler, even though there are other yellow ones. However, this bird appears under a new guise in the latest IOC list, with 'American' in front of the vernacular name, and an amended specific name. Is is now *Setophaga aestiva* (summer), the former name, *S. petechia*, having been recycled for another relative. The male Hooded Warbler which I saw in Florida was an easy one, because the black hood was so obvious, though you would need the scientific name *S. citrina* to help with the yellow-green female. I did once have some trouble identifying a flock of Chestnut-sided and Bay-breasted Warblers in Nova Scotia, since the dappled woodland light changed the browns into another version of black. In that case there is little help in the fact that *S. castanea* (chestnut-coloured) belongs to the 'wrong' warbler. Chestnut-sided has a purely geographical name, *S. pensylvanica.*

CHESTNUT-SIDED WARBLER

Clearly I have neither the right nor the opportunity to rearrange those which bother me, but having seen a Black-throated Blue Warbler in Nova Scotia, I would happily settle for Blue Warbler or even Blue Blackthroat: the latter form (and Green Blackthroat) would fit the same mould as Common Yellowthroat. As for the jazzy little Black-and-white job, my encounters with that on several occasions in both Florida and Canada fill me with the longing for less banality in its name. After all, the bird has a special magic, because it is the only warbler that behaves like a nuthatch in running down a tree or branch. 'Black-and-white' could so easily mean 'blotched' or 'pied', but in fact it wears jazzy 'go-faster' black streaks on a white ground. Surely we could do better for the handsome little bird with such a distinctive plumage and jizz?

As for Prothonotary Warbler, *Protonotaria citrea*, another yellow bird, I find that the more I consider that ridiculous name the more I like it. It appears pedantic, yet is eccentric and original. It belongs in the same category as the African Secretary Bird, not that the two show even the vaguest resemblance or relationship, but because someone with a sense of humour saw both as resembling a human type. The Secretary Bird certainly resembles a strutting clerk with a quill-pen behind his ear, but the Prothonotary is far less obvious as the golden-robed Byzantine lawyer. Obscure, certainly, a mouthful for sure, but worthy of being left alone.

I have seen Blackburnian Warbler, *Setophaga fusca*, on passage through Trinidad and in its breeding territory in Cape Breton. It was named for the English naturalist, Anna Blackburne, whose brother sent specimens to her from America. Mrs Blackburne was a contemporary and correspondent of both Linnaeus and Pennant. The dedication to a female amateur scientist is certainly unusual for the eighteenth century, but at least she had social status on her side: by contrast, the skills and knowledge of working-class Mary Anning as a fossil collector at Lyme Regis during the nineteenth century were exploited but largely unacknowledged by the scientific establishment. The fact that this beautiful species of wood-warbler acquired the dark word *fusca* is also little short of a scandal, as my singing male was showing off a gorgeous orange and black head.

It was in that in same patch of forest that I discovered why Common Yellowthroat had the scientific name *Geothlypis trichas*, which describes it as a ground-hugging thrush-like bird. Our male was foraging low in dense wild rhododendron and birch scrub, and it was only his intermittent song which gave him away as he flitted from bush to bush.

Ovenbird, *Seiurus aurocapilla*, shares its skulking behaviour, but takes its vernacular name from the domed grass nest which it creates on the ground. I was lucky enough to have a nest pointed out to me during a walk with a local group in Nova Scotia, but the young had already fledged. *Seiurus* means 'wagtail', while the specific name speaks of the golden-orange crown-stripe worn by the bird. We heard its distinctive song one summer, by a lake in Nova Scotia, but were forced to conclude that it should be related to the Cetti's Warbler, since it shared with it both a strong voice and a cloak of invisibility. One day!

The two wood-warblers which bear the name 'waterthrush' belong to the genus *Parkesia*, which commemorates a twentieth-century American ornithologist. They take the vernacular name from their water-side habits and thrush-like plumage. Having seen several Northern Waterthrushes, *Parkesia noveboracencis*, on passage through Trinidad, I found that, through binoculars and without comparison species, they can look much larger than they really are. Pipits can be deceptive in that way too. As for that strange word *noveborancensis*, it borrows the Roman name Eboracum (for York) to coin a word meaning 'from New York'. Now that is table-knife turned into a screwdriver.

Palm Warbler, *Setophaga palmarum*, was a common winter resident when we were in Florida, which helped to explain why this bird, which breeds mostly in Canada, is so called. In their drab winter plumage the two races, the western brown form and the eastern yellow form, both occurred in Florida and I saw both regularly. But what a beautiful bird it turned out to be when I found the really showy version in a Nova Scotian summer. Another familiar species in Florida was the Yellow-rumped Warbler, *S. coronata* (crowned), which brought us some magical experiences in woodland, where we were more than once surrounded by large feeding flocks, and we have seen them in Canada, much smarter, during the breeding season. Of course, I must now refer to them more specifically as the Myrtle Warbler, since the two former races have now been split. Audubon's Warbler occupies the western side of the continent.

Another special woodland sighting came during a visit to Green Gables in Prince Edward Island, this time in Anne's 'Haunted Wood'. A beautiful male American Redstart, *Setophaga ruticilla*, made sure that we saw the showy rudder which gives the species its 'copycat' name, but with the added bonus of the orange wing-patches too.

And finally, a word of warning for those, like me, who own an older American field guide (or four!). In 2011 a complete review of the classification of this family was announced. The familiar generic name *Dendroica* disappeared, to be replaced by *Setophaga*, and a number of species were re-categorised. This is not the place to discuss those changes in detail. Suffice it to say that Worm-eating Warbler remains as it was.

LET'S GO FOSSICKING: AUSTRALIAN NAMES

If the word 'fossicking' is unfamiliar, bear with me. It is one of those words which surprised me when I first heard it in Australia. A generation or two ago, any Australian could be stuck with the nickname 'Digger'. That derived from the days of the Australian Gold Rush. Even today many Australians do a bit of casual prospecting, known as 'fossicking'. It does pay: I know one lady who wears a beautiful emerald which her husband found shining in a cow's footprint in Queensland, and another who owns a beautiful opal mined by a friend with a hobby-mine. So this section is about my delving into some of the gems of Australian bird names. Apart from those, I also turn up a few treasured memories.

Looking for the native names

The obvious thing to look for among the names of Australian birds is evidence of those originating from Aboriginal roots. In such a massive continent, of course – we have to remember that Europe would fit into it with some to spare –there were some 260 languages and dialects among the native peoples who lived there before the arrival of English-speaking settlers. During my first visit, I started to wonder about some of the unfamiliar forms.

I soon discovered the name 'gerygone'. The Western Gerygone, *Gerygone fusca*, was not so much a gem as a sort of greyish-fawn warbler-like bird sharing a bush with an Eastern Yellow Robin in the Botanic Gardens in Canberra. When I located the curious name in the field guide, I wondered how it was pronounced and assumed that it might be linked to an Aboriginal language. That was because I hadn't spotted the Greek, *gērugonus*, lurking behind the name: 'a song that echoes back'. What a wonderful idea! However, it seems decidedly strange that a species which is found widely across the continent has the tag 'Western'.

The Emu looked a better bet. What is more Australian than its national bird, and a most un-English looking word to boot? Well, Macdonald removes any illusion by tracing the word from the Arabic for Ostrich, *Nā-amah*, which travelled with the Moors to the Iberian Peninsula to become a word for the Crane. The second syllable alone was then corrupted into Portuguese, before travelling with their explorers and traders to the East Indies, where it was applied to the Cassowary. Later the name was transferred by the Dutch to the Emu of New Holland.

So what about cassowary? That looked a good bet. No, it is Malay, or more probably Papuan, according to Macdonald, in which language it means 'horn-headed', a reference to its bony head-ornament. That begins to get a little closer, at least (which is more than we did when we searched for a Southern Cassowary in Queensland). *Casuarius casuarius* simply Latinises the original word.

Given that the word Emu metamorphosed from the Portuguese for Crane, it is time to consider a gem found in one of my favourite bird references, *Birds of Australia's Top End* by Denise Lawungkurr Goodfellow, an American who settled in the Northern Territory and was adopted into the Kuwinkju tribe. While the book is extremely useful, the descriptions, local hints and touches of humour make it worth reading for its novelty: how many bird guides suggest that 'you can try lying on your back and kicking your legs in the air' to attract a bustard? But she also has a wonderful paragraph about the origins of the name for Brolga Crane:

According to my Gunyok (sister-in-law), Brolga was named for a promiscuous young woman who loved to dance. When she would not give up her wayward lifestyle she turned into a beautiful dancing bird. A Roman, Aelian (AD 170–235), advocated cranes' brains as an aphrodisiac. Viagra is probably more effective and a whole lot cheaper – taking into account fines one risks for killing protected species.

So there we have the sort of evidence we need – plus an irreverent and irrelevant aside. If that commentary leaves you a little red-faced, so is *Grus rubicundus.*

Another species, Sarus Crane, *Grus antigone,* occurs in northern Queensland. This species is also found in Asia, as evidenced by the name Sarus, which is Hindi, and derives from the Sanskrit for a 'lake-bird'. As for the *antigone* element, there are two possibilities. The more well-known is the heroine of Sophocles and Jean Anouilh. That tragic daughter of Oedipus of Thebes hanged herself, which leads to a possible connection to the red neck of the Sarus Crane. That seems too facile an explanation. Jobling links the name to a far less well-known Antigone, the daughter of a king of Troy, whose legend is one of sacrilegious vanity and a vengeful goddess, Hera, who changed her into a stork. In another legend Hera changes a Pygmy princess, Gerana, into a crane for the same reasons. Jobling suggests that Linnaeus confused the two stories – and that seems to be the more plausible of the explanations. However, should you be lucky enough to need to separate the two species, the Sarus Crane does indeed have the red descending into the neck.

The link with India is repeated further. On each of my trips to Australia I have searched for the Pacific Baza, *Avecida subcristata,* fascinated by the images of a raptor with a pointed crest and by its strange name – but the bird continues to elude me, and its name turns out not to be Aboriginal after all. It seems that this is a name derived from a falconers' word, *baz,* which refers to a goshawk species in India and to a Peregrine in Russia. The scientific names describe it as a 'slightly crested bird-killer'. But there is at least one bird of prey, the Brown Falcon, *Falco berigora,* that does have an Aboriginal name – though only in the scientific form.

BROWN FALCON

I also discovered a similar, but apparently rare, usage of a word of probable Aboriginal origin in the specific name of the Scarlet Robin, *Petroica boodang*. My sighting of its Tasmanian subspecies should also bear the trinomen *leggei*, after the nineteenth-century co-founder of the RAOU, Lieutenant-Colonel W. V. Legge.

So where were the other Aboriginal names lurking, I wondered? The word 'cockatoo' looked like a possibility, so that was where I looked next, only to discover that it was another false trail. Having watched a pair of Yellow-tailed Black Cockatoos tear chunks of wood from the base of a eucalypt to get at grubs, I can well believe that *Cacatua* (Cockatoo), originates from an Indonesian dialect form, *Kakatuwah*, meaning 'vice' (as in the tool rather than as in bad behaviour, though both might have applied to the destruction wrought to the base of that tree). However, that generic name applies more correctly to the white cockatoos. Most of the black species belong to the genus *Calyptorhynchus*, which means 'covered bill', presumably in reference to the narrow cere.

YELLOW-TAILED BLACK COCKATOO

We found Cockatiels in the Northern Territory, and this is a name of similar origins: the OED suggests that it is from the Dutch *kaketielje*; Macdonald links it to the Portuguese *cacatilha* or *cacatehlo*; and Merriam-Webster states firmly that the Dutch form derives from the Portuguese. This would be quite logical, since Portugal was trading in the region from an earlier date. The suggested root links to 'cockatoo' and to its Indonesian origins, but as a diminutive form meaning 'little cockatoo'. The resemblance which underlines that name has only been recently confirmed biologically, and the species placed in the Cacatuidae. They have the scientific name *Nymphicus hollandicus*, the 'nymph' element reflecting their small and delicate appearance, while the specific name is a shorthand reference to New Holland (Australia).

It is in the realm of the cockatoos that I at last find another gem of native origin, with the Gang-gang, *Callocephalon fimbriatum*. That vernacular form is indeed an Aboriginal name, and one apparently adopted by early settlers. Given the bird's limited southeastern range, the original name would have come from that corner of the country. We saw them by the Hawkesbury River and in Canberra. The clumsy-looking scientific form refers to the 'beautiful fringe-crested head' of the male.

Another name which originates from the Sydney area may well have meant 'cockatoo'. Corella, Macdonald suggests, probably originates as a native word, *ca-rall*. Today it is applied to three species. We saw Long-billed Corellas near the Hawkesbury River and one of the races of Little Corella in the Northern Territory.

The cockatoos hold yet another gem. The Yuwaalaraay people, whose homeland is in the north of modern New South Wales, provided the word 'Galah'. Macdonald points out that it also had alternative forms such as *gila*, *gulah*, *gala* and *gillar*, all meaning 'bird'. The word happens to sit alongside 'drongo' as a word of mild insult, meaning someone who is loud-mouthed through drink; a fool; or a talkative woman, according to the target. The Galah, a beautiful pink and grey cockatoo, enjoys a scientific name which extols one of its less striking features, but twice, once in Greek and once in Latin: *Eolophus roseicapilla* is a 'dawn-crested, pink-headed' bird. In spite of its beauty, it became a noisy and persistent crop pest, so it tended to upset the farming settlers, who were clearly immune to its beauty. I never saw the birds in that mode, and they seemed to be tolerated in small numbers on the NSW farm which we visited. We also found them nesting in the Botanic Gardens in Canberra, in urban parks, and in the arid outback of Kakadu, so they are widespread and very adaptable.

GALAH

John Gould recorded that his 'Warbling Grass Parakeet' was known by the natives of the Liverpool Plains as the *Betcherrygah*, while another Aboriginal dialect had it as the *Gigerrigaa*. Budgerigar is a fairly recognisable derivative of either one. Macdonald points out that the RAOU preferred the native word over Gould's form in 1926, but in the spelling *budgerygah*, adopting the current form only in 1978. Their scientific name is *Melopsittacus undulatus*, meaning 'wavy song-parrot', the adjective referring to the markings on the wings and back. I have fond memories of a pet one I had as a lad, but, needless to say, none at all of the ones which avoided me in wide areas of the continent.

It seems entirely appropriate that several of the cockatoos should represent the original native names, since Australia is home to fourteen of the world's twenty-one species. The parrots clearly do less well in that respect, with only one of the forty-one Australian species retaining a link to native forms. In general, the parrots suffer from *déjà vu*, since the family was well known from America, Asia and Africa, long before the discovery of Australia. As a consequence 'book-names' prevail today, though one or two creations are worthy of note.

The lorikeets take their name from a Malay word *luri* (parrot) which is Anglicised as lory, and then used in a diminutive form for the Australian species, of which the multi-coloured Rainbow Lorikeet, *Trichoglossus moluccanus* (formerly *haematodus*) is the best known. The generic form there means 'hair-tongued', since the birds have a special appendage on the tongue to enable them to feed on nectar and pollen. The old specific name refers to the 'blood-red breast', while the Red-collared Lorikeet (formerly considered a northern subspecies) is *T. rubitorquis.*

The collective name of the six rosellas has a very unusual origin: they apparently take their name from a town near Sydney, known as Rose Hill, which is now part of Paramatta. From this, the form Rose-hillers led to the current name, via the species now known as the Eastern Rosella. That explanation really does come over like an Aussie leg-pull, but I have it on good authority, from Macdonald. Their generic name, *Platycercus,* means broad-tailed.

The Blue Bonnet is a prettily named species, an old folksy form acknowledged by Gould and which seems to have escaped the reviews mentioned by Macdonald. Appropriately, this particular gem, in its guise as *Northiella haematogaster,* bears the name of Australian jeweller-cum-ornithologist, A. J. North. The specific name refers to the 'blood-red belly-patch'. A recent split now requires the modified name Eastern Bluebonnet.

Double-eyed Fig Parrot is partly explicable for its habit of eating figs, but the curious element 'Double-eyed' seems like a strange joke, which is compounded by the fact that its scientific name is *Cyclopsitta diophthalma,* the Cyclops being the one-eyed monster of the Ancient Greeks and *diopthalmia* assuring us that the bird does in fact have two eyes. A single patch of colour on the forehead may account for the *Cyclops,* since the Greek *kuklos* means circle, and a sense of humour may well account for the rest. I have seen many of the other parrot species, but not this one. Of the three Australian races, one is considered endangered.

WONGA PIGEON

The pigeon family provided my next discovery. The Wonga Pigeon is a simple step away from a New South Wales Aboriginal dialect name, *Wonga-wonga.* I spotted this dumpy grey and white pigeon briefly on the ground early during my second visit and it led me a merry dance for the rest of the trip, since I regularly heard it uttering its repetitive notes from a tall tree near the Hawkesbury River. I eventually found it about two hours before I was due to leave, at a neck-breaking height in a massive gum tree. It was then that I realised why it has the specific name *melanoleuca* (black and white) since all that I could see were the black-spotted white underparts. It is otherwise predominantly grey and white. The bird has some right to be elusive, since it has a unique generic name, *Leucosarcia*, meaning 'white-fleshed', and comes with a recommendation from John Gould himself to the effect that it was ideal for the table.

The Wompoo Fruit Dove has always eluded me, but it too illustrates another rare instance of a name of native origin: this one clearly is an echoic form. This species shares with the very localised Banded Fruit Dove, which we saw in Kakadu, the genus *Ptilinopus* (feather-footed). That odd name sees to relate to feathering on the upper leg. The plumage of the former certainly earns it the alternative name Magnificent Fruit Dove, which is echoed in *P. magnificus*. 'Banded' is repeated in *P. cinctus*.

Significantly, though, the names of the remaining twenty-three Australian species of pigeon and dove show only the influence of the settlers and ornithologists. For example, Brown Cuckoo-Dove is a brown dove which has the shape of a cuckoo; Crested Pigeon has a pointed crest; Common Bronzewing is the commonest of the pigeons with scintillating bronze-coloured wings. Only Topknot Pigeon shows a touch of folksiness, and the problem there is that the name is often, erroneously, also applied to the Crested Pigeon.

The Southern Boobook looked quite promising as a native name: this owl apparently said '*boobook*' to the European settlers too, and the name stuck. It said '*boobook*' to me a few times, often at 2 a.m. outside my window, but flash-lamp sorties failed to find it. One particular sortie did, however, find the neighbours' Rottweiler-cross asleep on the veranda when I tripped over him. Fortunately he was a gentleman: we had been introduced, so he merely raised a slightly puzzled eyebrow. Naturally, however, with the commotion, the owl was off and away. Another Australian owl is called the Barking Owl, and it barked its way through the nights we spent under canvas in Kakadu. Mind you, I was not so sure of my facts when we saw a Dingo, the Australian wild dog, sneaking through the bush as we left camp, but our guide confirmed that we had heard the owl. We did eventually see a pair of these noisy owls in Darwin's Botanic Gardens, sitting, as good as gold and silent, while we were awake. Both these owls belong to the genus *Ninox*, which links *Nisus* (hawk) with *nox* (night).

Coming to the kingfisher family, I unearth with great ease a major gem. The Wiradhuri people of southeast Australia contributed the onomatopoeic *Guguburra*, which evolved into the form we now use in the name of the Laughing Kookaburra, *Dacelo novaeguineae*. This is another loud-mouth. The wonderful name Laughing Jackass was also given by the settlers to this remarkable giant of the kingfisher family to describe the near-lunatic crescendo of the bird's call, which is a fusion of braying donkey and hysterical laughter. However, the Jackass is a double-sided coin, since it is also linked to the French word *jacasser* (to cackle or chatter), which was, according to Baker (1978), and quoted by Macdonald, 'bestowed by French observers prior to 1798 on the Aus. birds'. In this element, there is a coincidence of meeting between the French and an unrelated English word which harmonises perfectly to give extra punch to the name. At the same time it recalls the sound roots of 'jack' in Jackdaw. For the record, the Blue-winged Kookaburra of the north earns the name Howling Jackass, with a performance which is hardly less manic!

As for the name *Dacelo*, it belongs in the same sad category as *Delichon*, in this case a rather flat anagram of *Alcedo*, the Latin for kingfisher, to which the kookaburras are closely related. Jobling quotes a marvellous outburst by Strickland in 1842, which is entertaining for its sheer pomposity. My favourite part of the quotation is that 'such verbal trifling as this is in very bad taste, and is especially calculated to bring science into contempt.' To me the cop-out of its creation is more redolent of a tired mind or of a lack of imagination, but crime against humanity it isn't. In any case the Ancient Greeks or Romans provided no precedent in the naming of Antipodean birds, so this is rather more excusable than the case of *Delichon*. As for the species name *novaeguineae*, this seems to be a piece of sloppy second-hand work by Sonnerat (1766), who had received a skin from Joseph Banks, Captain Cook's botanist. The confusion is compounded by the fact that Laughing Kookaburra does not inhabit New Guinea, while Blue-winged Kookaburra, *Dacelo leachii*, does. That sort of shoddiness seems to me somewhat more reprehensible than the creation of the word *Dacelo*.

From my experience, the Laughing Kookaburra starts its version of the dawn chorus about an hour before daylight. From a rural house by the Hawkesbury River, I estimated that the woodland held several families, thus ensuring that I never needed a wake-up call to be in time for a dawn walk. I well understood why one of their popular names is Bushman's Clock, or why Aboriginal legends have the bird announcing the lighting of the sky. It is true to say that the dawn chorus is somewhat different from what you are likely to hear in Europe. If you have got back to sleep after the Boobooks, you may have a short while before the Kookaburras start. Then it is the parrots and parakeets. And last, but not least, the Noisy Friarbirds join in. To be fair, there are some beautiful songs from the honeyeaters, robins and whistlers.

LAUGHING KOOKABURRA

One other member of the kingfisher family has a name which looks simple but is deceptively challenging: the Sacred Kingfisher, *Todiramphus sanctus*. There is little evidence in Macdonald to show how this name arose, other than the fact that the vernacular form was in evidence in Governor Arthur Phillip's *Voyage to Botany Bay*, written in 1789, thirty-eight years before the formal allocation of the scientific form. So to whom was it sacred? There is no suggestion of a link to local beliefs, so we have to look elsewhere to find that the widely distributed species was sacred to some Polynesian cultures and had undoubtedly been first named in that context, possibly even before Cook had arrived in Australia.

Another bird is linked to deities from further north. The odd-looking name Trumpet Manucode (a bird of paradise limited to Cape York) has roots in the Malay name *manuk dewata* (bird of the gods), which travels via the Latin *manucodia* to make the Anglicised form. In order to appreciate the 'trumpet' element of the name, you can do no better than consult Katrina van Grouw's stunning image of the de-constructed creature in *The Unfeathered Bird*, where the extraordinary coiled windpipe, responsible for the bird's resonating call, is revealed. It is reproduced here with Katrina's generous permission.

TRUMPET MANUCODE (KATRINA VAN GROUW)

In my pursuit of Aboriginal names, 'drongo' proved to be a red herring: the Australians use it as a harmless insult, but the fact that I saw a pair of Spangled Drongos in tropical Queensland should have given me a clue. I didn't know at the time that this was a member of a widespread Asian family which spills over into Australia, and that the name actually derives from a Malagasy word adopted by nineteenth-century naturalists. The distinctively shaped tail, which first caught my eye, explains the generic name *Dicrurus* (fork-tailed), while the 'spangling' is conveyed in the specific, *bracteatus*.

'Currawong' does, however, seem to fit the bill: the term relates to the call of the Pied Currawong. Several convincing onomatopoeic forms exist in several indigenous languages, for example *garrawan*, *gurawarun* and *kurrawang*. There are three species of these crow-like birds. We became most familiar with Pied Currawong in New South Wales, where they would come regularly to the woodland-garden feeders, and we saw both Grey Currawong and the endemic Black Currawong in Tasmania, all of them armed with an impressive bill.

The name 'pitta' was imported from the Indian subcontinent, from the Telugu language of the southeast: we encountered the Rainbow Pitta, *Pitta iris*, at Barramundi Creek in Kakadu, its multiple colours reflected in the spectrum of both its names.

Chowchilla, a bird of the Queensland rainforests, shows promise: it is an onomatopoeic name, but its linguistic origins seem to be unclear. Its close relative, the Logrunner, could not have a more down-to-earth name, since it describes the forest-floor habits of both species.

But, to conclude this section, there are three birds whose names derive from native words for the habitat in which they live.

The Mulga Parrot apparently favours the *mulga tree*, a species of acacia so-named by the Yuwaalaray and Kamilaroi peoples.

The curiously named Gibberbird, an ally of the honeyeaters, is an inhabitant of a terrain known elsewhere as 'desert pavement'. *Gibber* is a word of native origin which describes a compacted, stony environment of the east-central outback.

The language of the Wemba Wemba people of western Victoria gives the word *mallee*, a habitat of woody plains, and from that comes the name of the Malleefowl, *Leipoa ocellata*. Here, though, the scientific name is well worth a diversion. While the specific name describes plumage which is 'speckled with eyelets', the generic name, meaning 'egg-abandoner', reminds us that this is one of a family known as mound-builders, or megapodes. Those last two words are explained by the fact that the birds have evolved large feet with which to build nest-mounds of vegetation into which they lay their eggs, which are then incubated by the heat of fermentation. The Brushturkey and the Orange-footed Scrubfowl are the two other Australian representatives of the family. Scrubfowls foraged around the grounds of our hotel in Port Douglas, and they were again common in the parks of Darwin.

Given that there are over 750 species of bird in Australia, and that the Aboriginal cultures were closely intertwined with the natural world, the evidence of the handful of extant native names is comparatively slender, though there is some compensatory evidence of names which relate to cultures of the islands and continent to the north.

Plain in name only

When I first saw them, I couldn't resist the temptation to describe the Orange-footed Scrubfowl as 'pointy-headed moorhens', which I suppose is exactly how many settler-names began their life. Macdonald has several examples of names that have been supplanted. One of the most fetching is Kitty Lintal (sometimes preceded by the word 'Sweet'). The origin and meaning seems lost, though it is not inconceivable that there was an echo of Kitty Longtail, an old folk-name for Long-tailed Tit back in England. Alternatively, it seems equally reasonable to speculate that it may once have referred to a sweet-voiced girl of that name. The bird is now the Chirruping Wedgebill. In that case it is easy to tell which name was concocted by the farmer's daughters and which by the ornithologists. It is not always so easy to separate the two threads, so what follows is an exploration of another landscape in which the occasional gem shines out.

The very first native Australian species I ever saw was the wonderfully named Willie Wagtail – and that really is its formal name. There is an unusual homeliness in this bird's name and an echo of the 'old country' traditions of names like Robin Redbreast and Jenny Wren. The black and white bird, which belongs to a family of fantails, was sitting on a post at Sydney airport when I first saw it and, with one of the oldest-surviving European names, it seemed an appropriate symbol to start with.

Later that same day I saw my first Peewee – at least that was my sister's name for it. That turned out to be an original settler's name, possibly derived from Peewit, to describe the bird's call. That is a good example of a case where a popular name has outlived the imposition of a formal alternative. In fact the alternative is not particularly helpful, because the word Magpie-Lark is a parallel settlers' creation which simply tells us that it reminded them of two birds from back home, the Magpie in colour and the Lark in size. The generic name, *Grallina*, refers to its long legs (stilts), though the specific name *cyanoleuca* (dark blue and white) rather overstates the blue element of the black.

MAGPIE-LARK

Many other names, some perhaps created by the settlers, are no-nonsense and useful: catbird, riflebird, bellbird and whipbird all tell us about the sounds the birds make: I recall that my first encounter with a hidden Whipbird gave instant recognition. Butcherbird tells about what the bird does. Weebill and bristlebird tell us about their features (a small beak and facial bristles respectively). Spinifexbird tells us about the habitat of tufted *Spinifex* grass.

Pilotbird is a little less obvious, but the name was apparently given for its reputation for guiding lyrebirds. When we heard one singing in the Blue Mountains, it was over the lip of a precipitous cliff, so I was not in the least inclined to follow it.

Lyrebird is a little more romantic, but the impressive tail-feathers are just a bit exceptional and shaped to recall the ancient harp which gives the name. I still haven't forgiven the one that we could hear singing, but scuttled away at the slightest crunch of a dry leaf, so I refuse to dwell on its beauty. The song was a cracker, though, even if there were no chainsaws, as in that iconic sequence in Attenborough's film.

The curiously named Dollarbird, *Eurystomus orientalis,* is a summer migrant to Australia, and a member of the roller family. Its generic name means 'wide-mouthed'. The vernacular name looks deceptively plain and simple, yet it is something of a challenge. It was given by the early settlers for two distinct white spots on the wings, which fact would have little interest had it not been that the currency of the colony was then in Australian pounds. However, the word 'dollar' had long been in use in English slang, having derived from the Flemish *daler* as early as the sixteenth century. Indeed, it was quite normal in my young, pre-decimal, days to refer to the half-crown coin as 'half a dollar'. The British 'crown', a large silver coin worth five shillings (quarter of a pound sterling) had resulted from the amalgamation of the English coin with the Scottish dollar at the Act of Union in 1707. So the silver crown was the dollar in question, and not the much later Australian dollar.

Honeyeaters

The commonsense approach certainly applies in the naming of the honeyeater family, since that is what a great many of them do: they feed on nectar and pollen, some of them following a nomadic life in pursuit of a sequence of flowering eucalypts, banksias and the like. This family has always fascinated me, as this is the one which I feel exemplifies Australian passerines more than any other. Certainly, from a British perspective, it is one of the most overlooked of bird families. Only one monograph has been published, and it is long out of print and unobtainable. I have always felt that Darwin might have found them of similar interest to the Galapagos finches, but *The Voyage of the Beagle* makes little comment on Australian bird life. Having now seen half of the seventy or so Australian representatives of the family, I should like to know them better.

I earlier mentioned the dawn chorus and the Noisy Friarbirds. The friarbirds are a subset of the honeyeater family. There are three other Australian species: Helmeted, Silver-crowned and Little Friarbird, all of which I saw in the Northern Territory. They earn their name from the strange 'hood' formed by the skin of a relatively naked head. In fact one of their older names is Leatherhead. This feature seems to prevent head-feathers from being gummed up by nectar or pollen from the flowers on which they feed. This effect is most exaggerated in the three larger species. The adaptation is not unlike that of vultures, whose head and neck-feathers would be clogged with gore. And like them, the Friarbirds do not win beauty competitions. Not a face to kiss, one might say, yet that is exactly what their generic name *Philemon* invites in the meaning 'affectionate' or 'kissing', an idea relating originally to their call.

The generic name of the closely related wattlebirds is a little more to the point, since *Anthochaera* means that they 'enjoy blossoms'. As for the vernacular name, the fact that Australia has many species of tree known as 'wattles' is a red herring: the birds derive their names from small decorative flaps of skin on the neck, which are related to the sort of wattles seen in a domestic cockerel or turkey. The naming of these birds is interesting, because of a certain lack of logic. In names like Swallow-tailed Kite and Great Spotted Woodpecker it is obvious that the punctuation is precise and purposeful: the kite has a swallow-like tail and the spotted woodpecker has smaller cousins, not large spots. All four wattlebirds are brown. With Red Wattlebird and Yellow Wattlebird, it is the colour of the wattles which counts, so Red-wattled Bird and Yellow-wattled Bird would make better sense in principle. However, their two smaller relatives have no wattles at all, so the whole thing is best left alone. Though I have seen a great many of the Red and Little Wattlebirds, my one encounter with Yellow Wattlebird, a Tasmanian endemic, was with a juvenile calling for food and identified by its tail-pattern rather than by its wattles. One-off sightings like that bird become associated forever with a place, so it belongs for me with the Tasman Blowhole, with a Bandicoot which we also saw there, and with some friendly American birders who came to help us peer into a dense bush to find the caller.

RED WATTLEBIRD

A third group within the honeyeater family contains the four miners, whose generic name, *Manorina*, has nothing to do with the vernacular form and everything to do with 'narrow nostrils'. The Noisy Miner is another woodland species which loves to be near dwellings, as we discovered at the Hawkesbury River. On my first visit there, I was intrigued by a conversation with an Australian who was intent on comparing 'our miners' and the introduced ones. By that he meant the Noisy Miner and the Common Myna. I wondered whether the former spelling referred to the habits of the bird, but I later saw that these are canopy-foragers and build a cup-shaped nest in a bush. They do resemble the Asian myna family, and the spelling had simply been Anglicised, so they have nothing to do with the 'diggers'. The common origin of the two words (as well as the spellings mina and mynah,) is the Hindi word *mainā*. The Common Myna, *Acridotheres tristis*, was introduced in South Australia and is now a widespread pest, which is ironic for a species bearing a generic name meaning 'locust-eater'. Like the introduced Common Starling, it is 'notifiable' as a pernicious pest in the Northern Territory.

NOISY MINER

There are many others in the family which bear the name 'honeyeater', most of them preceded by 'book-name' variations on the colours white, brown, yellow and the like – and they are tricky indeed. There are exceptions, such as the New Holland Honeyeater, which I mentioned earlier. However, the name of one I saw near the Hawkesbury River struck me instantly: the bird was scarlet, black and white, so I didn't have to struggle to recall that there was one called the Scarlet Honeyeater. What a beauty! Then in the sunlight of one dawn sortie I watched two males sparring in a territorial dispute. It was one of the most magical moments possible.

Few of the names are that helpful, though Blue-faced Honeyeater is unmistakable for that striking feature. You might in fact be forgiven for finding some of the forms just a little confusing and hard on the memory. There are, for example, White-lined, White-gaped, White-eared, White-plumed, White-throated, White-naped, White-streaked, White-fronted and White-cheeked honeyeaters. Naturally enough, other colours feature strongly, with yellow coming a good second. Of over fifty species bearing the name 'honeyeater', most have similarly utilitarian names. In truth, some of the differences are extremely subtle and it is hard to imagine a system in which anyone other than the ornithologists would have named every single species. But there are some memorable names too, such as Spinebill (Eastern and Western), with their long decurved bills; and Crescent Honeyeater, with two distinctive black crescents on the breast. Only two are named after people: Lewin's after J. W. Lewin, an early settler, artist and ornithologist (a species which I saw regularly on my last visit); and Macleay's, after William Macleay, one-time president of the Linnean Society, NSW. Two more relate to habitat: Mangrove Honeyeater and Eungella Honeyeater, the latter taking its name from a small area of Queensland rainforest. Suffice it to say that this large family includes well over a dozen generic names, which would require far too much space to discuss and which in general are as mundane as the vernacular forms.

BLUE-FACED HONEYEATER

WHITE-GAPED HONEYEATER

Bright and early

I found that the trick of seeing these difficult birds was to rise before dawn, and sit quietly near water. I soon discovered that a good seat was useful: I once made the mistake of sitting on a bank occupied by bull-ants, the sort of error you make only once. Arriving before dawn at a water-hole in the Northern Territory, we saw several new honeyeaters, and saw finches better than at any other time. The stars of the show were the Gouldian Finches (John Gould's own). When I managed to get two red-faced birds into the same frame, Christmas had come early! It is estimated that only a few thousand of them now remain in the wild – and we had a bush full of them. They were such an icon that the actual views felt almost unreal. Most adults are black-faced, and proportionately few have red faces, and rarely a yellow face occurs. Our sightings that morning also included Masked, Crimson and Long-tailed Finch, which could all be readily identified from those names, and the local race of Double-barred Finches showed up too, looking like a little football team in their striped shirts. Though called 'finches' for their resemblance to the European species, all those at the water-hole belonged to an unrelated family, the Estrildidae, a name which reflects a form used by Linnaeus for the African species Common Waxbill, *Estrilda astrild*.

GOULDIAN FINCH

Quiet sessions by the Hawkesbury River also brought me into close contact with a small flock of Red-browed Finches feeding there. Now that is one name that could do with being used in its alternative form: Red-browed Firetail Finch tells the whole story. That same location provided me with unforgettable views of Grey Fantail, a small bird which spreads its tail into a dramatic fan for display. I discovered the magic of their hawking flights in pursuit of insects, when they flutter and manoeuvre with great skill, using the large surface of the tail to stop, twist and pursue with such grace and agility that I almost found myself believing in fairies. But of course the fairy-wrens had already seen to that. Whoever thought of calling one Superb and another Splendid did us no favours: the various species are difficult enough as it is.

Close by the fantail display, I found the beautiful gossamer-sewn pendant nest of the Striated Thornbill, yet another tiny bird which can easily be imagined from its name. A tiny thorn poked out of the entrance with the beady eye of a brown bird behind it. But, without a shadow of doubt, the songster of those scenes was the Australian Golden Whistler, whose name tells us succinctly that he is golden and that he sings a strong, sweet song. The simplicity of all those names is almost as helpful as the pictures in the handbook.

'Shrikes' and butcherbirds

That same site introduced me to a Crested Shriketit, and it was easy to see where the last element of the name came from: out of the corner of my eye I registered 'Great Tit', and then the penny dropped. As mentioned earlier, a slight hook on the bill accounts for the shrike element, as well as for the generic form *Falcunculus* (little falcon).

A 'little raven' is also introduced in *Coracina*, the generic name of the cuckooshrikes. Two of these, Black-faced and White-bellied, became familiar. They seem to take their name from a passing resemblance to the jizz of some cuckoos and to the hooked tip of the shrike's bill. Cicadabird belongs in this family, its name derived from a cicada-like trill. Such a song is also reflected in the name of the closely related White-winged and Varied Trillers, both of which we saw in the Northern Territory. These enjoy the generic name *Lalage*, which goes back to one of Hesychius's unidentified Greek birds.

Good views of two species of shrikethrush, Grey and Little, made me wonder why the shrike element was even considered: to my eye it seems to have no great relevance to either looks or habits. However, the association seems to be explained by the popular name Thick-head, which implies that they share a proportionately large-headed look with the shrikes (cf. Loggerhead). The generic name underlines the vernacular, since *Colluricincla* includes an old form *Collurio* (shrike) with *cincla* (thrush). I can certainly support the latter concept, since it was only the thrush-like song which we heard echoing around the rocks of Nourlangie, in Kakadu, which enabled us to record a near-encounter with the Sandstone Shrike-Thrush.

In Australia there are five species of butcherbird: these share some of the habits of our own shrikes, but belong to an entirely different group, the Cracticinae (from the Greek *kraktikos*, 'noisy'). Their bills are more substantial, and proportionately similar to those of the related Currawongs. Grey Butcherbirds were not averse to taking bread scraps at my sister's New South Wales house. During a torrential downpour one afternoon, one took refuge within three metres of me on the veranda, preferring my company to a drenching. Pied Butcherbird was seen at a Kakadu water-hole, and Black Butcherbird featured in Port Douglas, Queensland, where the begging calls of a young bird echoed around the hotel pool as the parents brought food. There is something special about floating on your back to watch a new species.

The Australian Magpie is a close relative: the birds are often bold, and their cheery warble forms the iconic 'sound of Australia' in many a soap opera or film. There are five visually different races, so it makes for an extra challenge to 'collect' these as you travel round.

The mud-nesters

The Corcoracidae take their name from a Greek form which was associated with the Jackdaw and Chough, and they are loosely related to the crows. These belong to the mud-nesters, so-called for building a cup of mud to form a nest.

We first encountered the White-winged Chough in a park in Canberra, not long after the first screening of David Attenborough's *Life of Birds* series, in which these birds featured strongly. To me it was like bumping into Attenborough himself. The White-winged Choughs travel in family groups and spend a lot of time foraging on the ground, when they appear to be wholly black: the white wing-patch is often not visible until they fly over an obstacle. I saw them in other places, but never found them to be as confiding as those in Canberra – and they were there again for me in the nearby Botanic Gardens when I went back several years later. Though these birds are not at all closely related to our Red-billed Chough, they are similar in size, shape and colour, though lacking the red bill and legs, and with a white patch on the wings. The name of the Chough was discussed earlier, but without mention of the fact that it is one of those nightmare words with which to test foreign spies for spelling and pronunciation. I was therefore intrigued when I later came across the following piece on a Canberra bird club website, by a wit called Ralph Reid. It is a beautiful commentary on the relationships between our spelling and pronunciation:

Choughs

One day while going out to plough
I noticed, sitting on a bough
A solitary white-winged chough
That preened and made its feathers flough
I thought that one was quite enough
But then I noticed flying through
The trees that there so thickly grough
Another one, and that made tough
I realised I was losing dough
Just standing doing nothing, sough
I turned around as if to gough
But as I did I gave a cough
That startled them, and they flew ough.

It says it all.

LOUSY JACK (APOSTLEBIRD)

The second species of this family is visually quite different, a smaller, scruffy, grey-brown bird, but with very similar habits. We sighted a flock during a visit to a farm in New South Wales. 'Oh!' said our host, 'They're Lousy Jacks'. One obvious explanation for that might be a healthy population of lice in the feathers, but Macdonald attributes the name to country slang where 'lousy' means 'petty' or 'fussy', while 'jack' means 'to cackle' (*jacasser* again). One alternative name is 'CWA birds', a name created by some non-PC joker because they reminded the men of an outing of the Country Women's Association. Another term, Grey

Jumpers, is almost banal after those names, since it relies on the birds' grey plumage and jumping habits. But it was the formal name, Apostlebirds, which I found intriguing. Like the White-winged Choughs, these birds live and forage in extended family groups, with adult young in attendance of new siblings. The reference in this name is to Christ's twelve Apostles, though the families may be smaller or larger than a dozen.

Other vernacular names

Nowadays, such 'folksy' names are hard to find in the modern field guide, but odd ones do survive: Jacky Winter is still the name of a small robin-like bird. The song may account for the 'Jacky' element (*jacasser*), but the original intention of the word 'Winter' seems already to have been lost by 1827, when naturalists Nicholas Vigors and Thomas Horsfield attempted to find an explanation. Was it a play on somebody's name, or was it perhaps a reference to distinctive snow-white of the outer tail-feathers? We shall never know, but the name is far preferable to the deadly-dull alternative of Brown Flycatcher, which now seems to have been set aside in favour of the older form. A close relative was called the Lemon-bellied Flycatcher when we saw it in Kakadu, but the last element has now been replaced by the delightful form 'Flyrobin', which presumably avoids confusion with the monarch flycatchers and other types. Apparently Jacky Winter also had the name Post-sitter, for self-evident reasons. Such names are representative of a situation in which settlers, often in isolated communities, created new names off the cuff. Memorable forms caught on, and often flourished, but most were eventually weeded out by the intervention of the ornithologists.

Apart from those already mentioned, a number of names contain a certain element of 'local colour' simply because the birds themselves are far-removed from those we know at home. It is the compound forms which sometimes seem exotic; 'rifle' is obvious and 'bird' is obvious, but put the two together and the name 'riflebird' suddenly becomes exotic from a British perspective. That suggests that an element of imagination has suddenly been sparked by the marriage of two ordinary words. I have already introduced a number of these, but there are one or two more which deserve further comment.

Of these, the bowerbirds often feature in televised documentaries because of the strangeness of their courtship rituals and their creation of decorated bowers. To actually see the bower of a Great Bowerbird was a huge thrill. The gawky brown bird itself was a bit mundane, but the stocky Satin Bowerbirds were a real treat visually, the males a shining purple-blue-black and the females in the most delicate shaded greens and browns. What did surprise me was to find flocks of the latter both feeding on the ground and in woodland canopy.

Figbirds, whose diet includes the fruit of the name, possibly catch the imagination because the fig itself is an exotic fruit to the British. They have become familiar to me, represented in both Queensland and Darwin by the yellow-bellied race, *flaviventris*. The Figbird is one of the 'Old World' orioles, which means that I did initially struggle to separate the greenish, streaked female from the other two Australian oriole species, which I discussed earlier.

At first sight, Mistletoebird strikes as a vaguely familiar name, but there is just enough oddness about it to warn that it will be different. And how! Not, as we might expect, a thrush-like brown job, but rather a vision in scarlet, black and white, a member of the tropical flowerpecker family. Not unsurprisingly the bird takes its name from the fact that it consumes fruits and berries, especially those of a mistletoe species. Our first awareness of the species was of a beautiful cup-shaped nest identified by a local guide on the Daintree River, but the sight of a stunning male singing in the evening sunlight at the top of a tree in Darwin was something extraordinary.

Add the familiar word 'silver' to the banal word 'eye', and once again you conjure up something exotic. In this case, Silvereye, *Zosterops lateralis*, is an Australian species of white-eye, a grey-green, warbler-like bird, with a marked eye-ring (hence *Zosterops*), and one which I first saw in Canberra. Tasmania has the nominate subspecies, with subtle shades of blue-grey, green and chestnut, the latter pronounced on the flanks to earn the name *lateralis*. I was glad to see one well at Strahan in the west of the island. It is worth a small digression here to note that two of its folk-names reflect total contrasts: Grape-eater tells us that it can be a bit of a pest, while Blightbird speaks of its capacity to devour harmful aphids. Storm-blown migratory birds of the nominate race colonised New Zealand in 1832, where the Maori name *Tauhou* still indicates it as a 'newcomer'.

Names such as Grasswren, Fernwren and Scrubwren immediately alert us to the fact that these are different from species at home, and in these cases 'wren' is used in a similar way to that discussed in the earlier section on European warblers – that is, as a generalisation for any small, brown, skulking bird. I once recall doing some work on my sister's vegetable garden, where the presence of White-browed Scrubwrens made me feel quite at home, since they were feeding under a hedge, just as Dunnocks do. (Mind you, I had been reminded to take considerable care in the garden, since a deadly Brown Snake had been seen among the vegetables a few days before.) We encountered two endemics of the same family at Cradle Mountain in Tasmania – Tasmanian Scrubwren and the Scrubtit – during a walk which mopped up several more such specialities. We were fortunate to find the area in one of its rare sunny moods, and the walk round the lake remains one of the most memorable we have ever undertaken.

If I were to choose a single Australian bird as the one not to be missed, it would undoubtedly be a species with one of the simplest of common names, the ugly, brown-grey Musk Duck, which gets its name from the male's habit of emitting musk in the breeding season. Its scientic name *Biziura lobata* speaks of a 'straw tail' and 'lobes', but nothing about that strange combination prepares you for the sight of it. Imagine, if you will, a large steam-rollered female scoter, a sort of aquatic skateboard of a duck, with a spiky tail and a leather purse dangling from its lower mandible, and you have something completely unique in the avian world. It is a reminder of Gondwanaland, of the profoundly ancient past of a continent which produced so many strange variants and left behind this jaw-dropping, unique branch of the tree of birds among so many other wonders.

***

In summary, it is sad to find that, in the current vernacular of Australian bird names, there remains limited evidence of the languages of the original Australians, a fact which does hint strongly at the cultural values of a sometimes aggressive colonial past, but also reflects the fact that there were no written forms for the many and varied native languages. It seems at first a little odd to find a number of other names which are clearly not English in origin, but which represent cultures from further north. Names like 'drongo' and 'cockatoo' had been adopted before Australia's colonisation and simply moved into Australian usage as other species of those families were identified around the colony. In the 240 or so years since the landing at Botany Bay, the English-language element of bird names has changed considerably. However, we have to remember that Thomas Pennant, for example, represents an equivalent milestone in the development of British names, and that a lot has happened to those since that time. What strikes me most strongly about Australian vernacular names is that most are fairly transparent, and offer useful information, yet the simple fact of compounding two quite simple words so often creates an exotic feel which is sufficient to alert the British birder to the fact that there is a lot of excitement in Australian birding – and plenty of pathless wildernesses too.

THE TRINIDAD TRAIL

Once again I want to follow a trail of themes through my discussion of the names of the birds of Trinidad.

Trinidad is an island just off the coast of Venezuela. In a cultural and sporting context Trinidadians ally themselves more with the Caribbean nations than with South America. Their national language is English, the language of the last colonial rulers, albeit in a strong local form which reflects both a complex ethnic mix and former rule by both Spain and France. However, the island's ecology and avifauna have a clear affinity with mainland South America, so this trip brought us into contact with many new species, some of which represented new families which we had not met in North America.

Not only were most of the birds new to me, but the colonial history of the Americas guaranteed that the names of many of the bird species found there would reflect a diversity of influences: Spanish, Portuguese, Dutch and French in particular would be interwoven with the English language. As with Australia and North America, I set out hoping to find also the influences of the languages of the original native peoples, and this time I found many strong traces, these deriving often from mainland South and Central America. One bonus was an element arising from a more recent Antillean culture.

Fortunately I have had more time to follow the linguistic trail than the seven days I spent in the field actually birding in Trinidad. If I haven't turned up all the available treasures, it is simply that time and space impose their own limitations.

The lure of the exotic

I rather expected that the Tropical Parula, *Setophaga pitiayumi*, would be showier than its Northern cousin, simply because of the word 'Tropical'. In fact it is normally a little brighter than the Northern Parula. So when I eventually saw one in Trinidad it was, in fact, an anticlimax: ours was a very dull immature individual. That strange-looking word, *pitiayumi*, was used by Vieillot in 1817 to respect its Paraguayan name, which, he explained, meant *'petite poitrine jaune'*, 'a little yellow breast', but ours was a dull little brownish-grey job. In a way I didn't mind, because 'Parula' still says subtle pastels to me.

Trinidad produced many more magical moments, of course. Having arrived at the Asa Wright Centre in the dark, we opened the door at dawn to look for our first new birds – and there they were: a Tropical Kingbird on a wire; a Blue-crowned Motmot on a tree; Bananaquits in the flower-bed; and a nesting Barred Antshrike in a bush by our steps. What a greeting! And what a bunch of names to start with!

BARRED ANTSHRIKE

There is something magical about things exotic, which clearly influenced such as the nineteenth-century Romantics, who aspired to recreating in their writing an atmosphere of 'local colour'. In a time long before film and television, they needed a technique to stimulate the imagination. By sowing exotic words into their writings they strove to recreate word-images which made the adventures of overseas explorers and colonists take on three dimensions. I am certainly not immune to a fascination with such things, and I really enjoyed discovering birds with names like 'jacamar', 'cacique', 'oropendola', 'macaw', 'toucan', and many more: the names were as colourful as the birds, and that made them all the more exciting.

The concept of 'exotic' is a relative one, of course, and this was never brought home more clearly than when some of the younger generation of our family came to visit us in England from Canada. They had been here for only a short time when a Magpie appeared on our lawn. Wow! Their reaction was one of total amazement at the sight of this spectacular and wholly unfamiliar bird. Seen objectively, you have to agree that it is a striking bird. It was a little easier to understand their excitement when a Green Woodpecker came down later. The truth is that we can take our own common species too much for granted: in Europe we have some of the most beautiful ducks in the world, but we can so quickly treat Teal, Mallard, Wigeon and Shoveler as background while we search for less common species. It works in reverse too: how many of us have encountered the situation abroad when the locals wonder why their common species should be so important to an eccentric foreigner with binoculars or a camera?

But let me return to some of those Trinidad birds, and to that paintbox of wonderful names.

'Jacamar' is one of those convincing exotic words, derived from the native Tupi-Guarani form *Jacama-ciri*. The local species was Rufous-tailed Jacamar, *Galbula ruficauda*. The scientific name refers to a 'yellowish bird with a rufous tail', so there are no surprises there. I recall that I had quite a job locating this rather showy bird, since its shaded colours blended so well with the vegetation. It also helped to know that we were looking for something halfway between a kingfisher and a bee-eater, but with a touch of hummingbird, so if you want exotic...

'Motmot' is a far simpler concept, and is akin to the story of 'Boobook', since the name is related to the bird's two-note call, but this time it was an Aztec word adopted by Hernandez in 1651. The scientific genus *Momotus* warns that attention to detail matters. Willughby was guilty in 1678 of dropping the first *t* when he Latinised the word, though the missing letter may just make it a little easier to say. (In truth, I imagine him sitting there, twisting his tongue, and then exclaiming 'Half a mo!' as the obvious solution appears.) This bird graced the cover of my handbook, so it seemed entirely appropriate that we saw one as soon as we emerged into the dawn light. It being the nature of first impressions, we saw only one or two more during the week. But in this species lies a warning of the challenges facing any birder who tries to maintain an accurate list. There has recently been a five-way split of the Blue-crowned Motmot, *Momotus momota*, and this one is now the Trinidad Motmot, which gives the island a second endemic species. However, the irony of the name *Momotus bahamensis* must take some of the gilt off the gingerbread for Trinidadians. It appears that when Swainson first described this form in 1837 he erroneously gave the type-locality as the Bahamas. That was corrected, but the name remained. In any case, there are no motmots anywhere in the West Indies avifaunal region.

'Macaw' is a curious word, which the OED tells us has a Portuguese root of unknown origin when referring to the parrots. In the next entry it states that it is a Carib West-Indian word for palms of the genus *Acrocomia*, which include the Macaw-palm. An online dictionary suggests that it is also related to the Tupi word *macahuba*. Two things strike me immediately about the two definitions. Firstly, the Portuguese had a foothold in Brazil, the homeland of the Tupi, as early as 1500. Secondly, we watched Red-bellied Macaws, *Orthopsittaca manilatus*, eating quantities of palm nuts at Waller Field in Trinidad. Is it too obvious to suggest that the concept of Macaw-bird might well have sprung from such observations? The scientific name of our macaw speaks of a 'straight parrot with broad hands', which refers obscurely to the shape of the cere and the breadth of the wings.

As to the other parrot species, we saw Blue-headed Parrot, which has the curious name *Pionus menstruus*, the generic name meaning 'fat' and the specific referring to the blood-red under-tail coverts. A very odd name indeed. We also saw Orange-winged Amazon, *Amazona amazonica*, which was common at Asa Wright. Two tiny species also put in appearances. At Matura we saw the Lilac-tailed Parrotlet, *Touit batavicus*, which has another Tupi name (from *Toui eté*) and a reminder that the French naturalist Edme-Louis Daubenton (1732-1785) apparently confused Batavia (the old name of Jakarta, the capital of modern Indonesia) with Venezuela – an even wider miss than in the case of the Eastern Curlew. If that last bird was tiny, the Green-rumped Parrotlet, *Forpus passerinus*, is even tinier, as suggested by the 'sparrow' element of its name. It is in fact slightly smaller than the Great Tit. The genus name *Forpus* is of arbitrary origin and may come from the Greek *phoreo* (possess) plus *pous* (foot) to underline its dexterity with its feet. We saw that one on two occasions in open country.

The word 'toucan' is a little easier to trace than 'macaw'. The word derives from the Guarani *tucán,* which means 'bone-nose', an obvious reference to the large bill. We did see them occasionally, but less than we had hoped, sometimes high on the forest canopy below the veranda at the centre, and once in dense trees by the mountain road. This was Channel-billed Toucan, *Ramphastos vitellinus.* The first element of the scientific name derives from the Greek *rampē* (bill) and the second from the Latin for 'egg-yolk', to denote the rich colour of the breast. 'Channel-billed' appears to relate to a groove along the joint of the bill, as opposed to a clean joint or an overlap. The same form appears in the even more dramatic Channel-billed Cuckoo, *Scythrops novaehollandiae,* which we saw twice in the Top End of Australia (but in that case the generic name suggests a scowling face).

'Cacique' is also comfortingly authentic, in that it derives from a Caribbean word, *cazique,* for a native chief. Yellow-rumped Cacique, *Cacicus cela,* was our Trinidad species. Curiously the word *cela* is obscure, and Jobling speculates that it may be a Linnaeus shorthand for *kelainos,* the Greek for black. That would make some sense for the black birds with striking yellow markings which we saw flying into their pendulous grass nests high in a canopy, but it does seem that Linnaeus might have had a bit of an off day when he described that bird.

The oropendolas, like the caciques, are members of the family Icteridae. In Europe, the word *Oropéndola* is the Spanish for the Golden Oriole, and reflects in the name the bird's suspended nest. As in the English-language version, the word has been re-used in the Americas for similar, if unrelated birds. The spectacular bird which we saw often and well in Trinidad was the Crested Oropendola, *Psarocolius decumanus.* As if the erroneous oriole link was not enough confusion, the name *Psarocolius* combines the Greek words *psar* (Starling) with *koloios* (Jackdaw). Thankfully the addition of the Latin *decumanus* (large, as a Roman legionary might be) emphasises that this is a sizeable version of this curious mix-up. And indeed it is a curious-looking bird, chocolate and yellow, with an ungainly large ivory bill and a magnificent pendulous nest. With all of those features, surely someone could have given it some distinctive first-hand names, rather than that hotchpotch of second-hand cast-offs?

OROPEDOLA NESTS

The New World orioles were represented on our trip by the Yellow Oriole, *Icterus nigrogularis*, and the Moriche Oriole, *I. chrysocephalus*. *Icterus* is at the very root of the family name Icteridae, to which belong the caciques, oropendolas, grackles, cowbirds, blackbirds and meadowlarks of the Americas. The names of the two orioles mean 'black-throated' and 'golden-headed' respectively. However, it is the common name of the Moriche Oriole which is of greatest interest, since it tells us that the bird is associated with the widespread South American Moriche Palm, *Mauritia flexuosa*. The Spanish word *moriche* derives from the Tupi *muriti*. We saw just one Moriche Oriole, and appropriately in a palm tree along Manzanilla Beach. Interestingly, I discover that on the current IOC list this is now treated as a subspecies of the more widespread Epaulet Oriole, *Icterus cayanensis*, so called in English for its bright yellow shoulder-flash.

The concept of exotic is exemplified by the Trinidad Piping Guan, *Pipile pipile*, which was undoubtedly the most important species we saw during our visit. It is currently listed as 'critically endangered' due to illegal hunting and habitat destruction, with maybe as few as 200 left in the wild. Part of the problem may be explained by one of its popular names, Wild Turkey. Sadly it is one of only two Trinidadian endemic birds (with the motmot), yet protection and conservation measures appear to be frustratingly ineffective. According to Merriam-Webster, the word 'Guan', first known in 1743, had the original form *quam*, probably from *kwama* in Kuna, the Chibchan language of Panama. Trinidad's guan has the local name *Pawi*, which is a corruption of the Spanish *Pauji* (Peacock). Winer cites the alternative Trinidadian forms of *paoui*, *pauie* and *pauxi*, the last form coinciding with the scientific name of the Helmeted Curassow, *Pauxi pauxi* (which does not occur in Trinidad). Curiously, the name *pauji* is used for the related curassows in mainland Venezuela, while the Spanish for the guans is *pava*, meaning 'female Turkey'.

TRINIDAD PIPING-GUAN

The use of the suffix *–quit* in Bananaquit and Grassquit has a very distinctive Antillean flavour. It seems to denote a small bird rather than a species or family. The origins of the word seem to be obscure, and it may simply reflect the sound made by such small birds (compare 'tweet' as the origin of the name Twite). Bananaquit, *Coereba flaveola*, seemed to be fairly common in Trinidad and fairly bold too, since, as nectar eaters, they frequently commandeered the sugar-feeders on the balcony at Asa Wright. They have blue-grey upperparts and a bright yellow breast, the latter providing the *flaveola* part of the name. The Trinidad race is *luteola*, which implies 'saffron yellow'. *Coereba* is said to derive from a Tupi name, *Güirá coeriba*. It is a widespread species throughout tropical South and Central America and the West Indies, with a confusingly large number of races and a number of potential splits in the offing.

We saw two species of grassquit: Sooty Grassquit and Blue-black Grassquit, the latter being more abundant than any of the others of its type. There is apparently a good reason for this: Blue-black Grassquit does not sing prettily, so it is not trapped for the cage-bird trade, a fact which has rendered the singing species of this and other families so scarce on the island. It was certainly a bizarre sight to see grown men walking with a handbag-sized cage containing a bird, as they set out for an evening of song-bird contests over a drink. As for the Blue-black Grassquit, *Volatinia jacarina*, he replaces his song with a manakin-like display in which he hops up from the branch as if on a strong spring. To the locals he is Johnny Jump-up, surely one of the most charming of familiar names, and one which is echoed in the Latin and Tupi words of the scientific name, all of which tell us why he isn't in a cage with his cousins. We saw this display once or twice, and it was certainly a wonderfully comical performance. There are clearly times when actions speak louder than words, or when it pays to keep your mouth shut.

Similar performances were given by two other species which we saw well at the Asa Wright Centre: Golden-headed Manakin, *Dixiphia erythrocephala*, and White-bearded Manakin, *Manacus manacus*. The word 'manakin' derives from the Middle Dutch *mannekijn* and occurs in another spelling, as in Chestnut-breasted Mannikin, *Lonchura castanothorax*, a wholly unrelated finch which we saw in Australia. It denotes a little person and undoubtedly relates to the peculiar dances of the males at the lek. While the afternoon heat provoked siesta mode for some, I took myself to a corner of the forest where we had earlier seen the White-Bearded species lekking, and spent a glorious hour watching the ground-level arena and the fascinating hops, flutters and gyrations of one or more males. *Manacus* could easily have meant 'manic', but it is in fact just a dull Latinised form of the English name. (As for the Australian bird, *Lonchura* means lance-tailed, while the specific name matches the vernacular form.)

The French language comes into the name used widely for two species closely related to the kingbirds, the kiskadees, an echoic name based on a call resembling the French phrase *Qu'est-ce qu'il dit?* (What's he saying?). Great Kiskadee, *Pitangus sulphuratus*, takes its generic name from a Tupi word for 'flycatcher', and its specific name from its 'sulphur yellow' breast. It was common at Asa Wright, though Boat-billed Flycatcher, *Megarynchus pitangua*, was confusingly similar and I almost overlooked that species as a result. The resemblance is acknowledged in the specific name, while the generic speaks of its 'large bill'

I recall that on that same afternoon I also located the local Common Potoo, *Nyctibius griseus*. The strange name 'potoo' derives from the nocturnal call of the bird, while the scientific name rather unimaginatively tells us, in a mixture of Greek and Latin, that it is a 'grey, night-feeding bird', a description which would suit just about any of its nightjar cousins. This was in fact a brown-morph bird, and we had previously seen a grey-morph individual in Caroni Swamp, which had pleased my wife immensely, since she had set her heart on seeing one. When I found the second bird, Mary was having a siesta, as we were due to be out late that evening. The afternoon was hot and the Potoo was sitting drowsily in a tree by the path below the Asa Wright Centre. It was doing nothing but pretending to be a branch. I took a series of very clear photos, and what emerged was that a bird which appeared to be totally still was in fact moving subtly and keeping a hooded eye on me. This is a fascinating creature, which relies on its camouflage to perch on open snags. And it had to be confident of the ploy, since we saw eighteen raptor species during that trip, including a nearby Ornate Hawk-Eagle.

COMMON POTOO

The Pauraque, *Nyctidromus albicollis*, is more of a 'standard' nightjar, and has a name which is also used in North America, where its range just seeps over the Mexican border. Curiously it appears in Sibley as the Common Pauraque, even though no other bird bears the name. The root of the word is Spanish, but probably based on an echoic Native American form. The generic form, *Nyctidromus* (night-runner), refers to the bird's rapid nocturnal flight, while *albicollis* refers to a white throat-patch. The species has a number of popular names in Spanish, of which *Tapacamino* (road-coverer) is a direct reflection of the way we saw both Pauraque and White-tailed Nightjar on the former runways of a disused airfield, where the retained warmth of the concrete attracted insects in the cool of the night.

The name of another loosely related species, the Oilbird, *Steatornis caripensis*, has less of an exotic ring to it, but it was an even more wonderful bird to see. This is a nocturnal fruit-eating bird, the only one of its sort, which lives in caves. The Asa Wright Centre has its own colony. Another very warm afternoon, and a steeply undulating walk, took us to the cave where we had glimpses of the birds on their nests. One might expect the oil to be of the sort ejected defensively by Fulmars, but rather it was of the sort produced by well-fattened young which could be collected and used for fuel-oil, hence the Greek word *Steatornis*, 'a fat bird'. This use of the birds was reported by Humboldt (1767–1835). The species is distributed in cave colonies, so the Caripe Caverns of Venezuela donate the second part of the name. The tale told by our local guide of the former tradition of a beheaded young bird, equipped with a rush wick, and used as a candle to light the caves during collection, was a graphic enough illustration of the potency of the oil. I mentioned earlier, when discussing the word 'puffin', that the collection of young seabirds from the nesting-cliffs for use as fuel and food was once the norm in some parts of Britain. Such subsistence collection was an absolute necessity in remote communities like St Kilda. We heard a similar story, when we visited the Bruny Island colony, of the former use of young Short-tailed Shearwaters in

Tasmania. In order to survive their stranding in the Antarctic, the Shackleton expedition cooked penguins in fires fuelled with penguin oil. Today, however, the Oilbirds enjoy a highly protected status, so the caves are only visited once a week and only a brief glimpse is allowed and in low torchlight. The birds' nocturnal habits, and noises, like those of the owls, caused some unease. This led to two alternative local names, the Spanish *Guacharo* (wailer) and the French *Diablotin* (little devil/sprite).

Oil returns again, but this time as Pitch-oil, one of the local names quoted by Richard ffrench for the Greyish Saltator, *Saltator coerulescens*. *Saltator* is the Spanish for dancer, a word inspired by the heavy hopping movements of *S. maximus*, Buff-throated Saltator. The concept of 'bluish' (*coerulescens*) rather exaggerates the dull slate-grey tones of the male Greyish Saltator.

The tanagers too are associated with dance. The word originates in Tupi as *Tangara*, meaning 'dancer', before translating into the Portuguese form *Tanagra*. The adjective *coerulescens* might certainly suit one of my favourites, the Blue-grey Tanager, *Thraupis episcopus*, which we saw regularly. However, the powder-blue colour is conveyed inaccurately by the word *episcopus*, a usage derived from the bishop's purple. The form *Thraupis* has been preferred in recent years for some of this family, but is much older than the discovery of the New World, having been trawled from Aristotle's term for a small finch-like bird. In fact we saw all but one of Trinidad's tanager species, including the spectacular Turquoise Tanager, *Tangara mexicana*, and the three related honeycreepers. With this family there is a full paintbox of some of the most spectacularly coloured birds in the world.

Classical drift

Inevitably, the influence of the Classical languages has come to dominate some of the names, and that is true of the gorgeous Blue Dacnis, *Dacnis cayana*. This was a real stunner, even among such showy birds. The Greek *daknis* was originally used for an unidentified Egyptian bird.

Another Greek word, *Euphonia*, derives from a musical term for a beautiful song. That is not true of the feeble song of the Trinidad Euphonia, *Euphonia trinitatis*, which, in spite of its name, has a range extending deep into Venezuela. Violaceous Euphonia, *E. violacea*, does have a far better song, though the 'violet-blue' of the plumage is quite similar. The word *trinitatis* is the Latinised version of the Spanish *Trinidad*, the name given to the island by Columbus, who, in 1497, saw the three main peaks as a symbol of the Holy Trinity. The current IOC list now places the euphonia species in the Fringillidae, and close to the European finches.

We found all three of Trinidad's beautiful trogon species, all quiet, rotund birds with striking plumages. But once again it is Greek which prevails in that word, since the word *trōgōn* means fruit-eating and was used by Pliny. Their names, both English and scientific, hold no other mysteries: Violaceous is *T. violacea*; Collared is *T. collaris*; and White-tailed is *T. viridis* (green).

And while we are still with Greek, the Black-tailed Tityra, *Tityra cayana*, a noisy bird, takes its name from the Greek satyrs, who caroused with Pan and Dionysus. Greek also appeared in the name of our 'bogie-bird', the Streaked Xenops, *Xenops rutilans* (bright rufous). Being a canopy gleaner, it was as elusive as any bird could be and our group leader was determined to find one, so much so that we jokingly started to refer to it as the *X-nops*, a sort of teasing non-bird. *Xenos* is the Greek for 'strange' and *–ōps* means face. However, when it eventually did appear, just a few hours before we left, I recall it as a disappearing rear end, as so often is the case with a bird which someone else finds.

We witnessed the performance of the Bearded Bellbird, *Procnias averano*, on our first morning in Trinidad. Its English names tell us a lot about its appearance and song, which both feature in a form of lek display. The name *Procnias* derives from Procne, the princess of Greek myth who transformed into a swallow. The connection escaped my eye entirely, because this was a bird with a beard like Zeus himself. Indeed the subspecies name is *carnobarba* (fleshy beard), which describes the mix of feathers and wattles which the male displays. The species name *averano* is more helpful in linking this unlikely bird to the swallows, since it was adapted by Buffon from the Brazilian Portuguese name, *ave de verão* (bird of the summer), which is, of course, how we consider the Swallow. The nominate species is reported to be endangered in Brazil, where, like Trinidad's grassquits, it is persecuted for the cage-bird trade. I can think of few areas of human activity as paradoxical as caging something which you apparently admire. There is no more frustrating picture than that depicted in Tony Juniper's book about *Spix's Macaw*, which illustrates how ego, greed and acquisitiveness can strangle the very existence of what people purport to enjoy and value.

FERRUGINOUS PYGMY OWL

I wrote earlier of Australia's noisy owls, but I have never met a noisier owl than the Ferruginous Pygmy Owl, *Glaucidium brasilianum*. Because the climate is so kind, our room at the Asa Wright Centre had no glass in the windows. Grilles and fly-screen are not soundproof, so our first night there was enlivened by one of these tiny owls sitting close by and doing a very good imitation of a squeaky water-pump. I counted one sequence of twenty-eight continuous calls (I was too excited to sleep in any case). The metallic nature of the call is appropriate to a bird whose first name implies that it is the colour of rusty ferrous metal. At a practical level we saw a lot of small birds because of this call: our local guide imitated it perfectly, with the result that small birds bundled out of the bushes, intent on mobbing the 'owl'. The scientific name speaks simply of a 'Brazilian little owl'.

The hummingbirds

With exotic in mind, the most obvious birds to demand some close attention are the hummingbirds, which are a speciality of the Americas. On the face value, this is one of those plain English names needing no great explanation. It relates to the buzz of the rapidly whirring wings, which, Richard ffrench points out, have been calculated to be up to eighty beats per second in some species, but there is much more to the naming of 'hummers' than meets the eye. To start with there are no fewer than 342 species in the Americas.

I met my first members of this large family in most unexpected circumstances, since I was in Nova Scotia at the time. I have since become used to them as a garden bird at various homes we visit in Eastern Canada. From the perspective of the average British birder it comes as a bit of a surprise to realise that Ruby-throated Hummingbirds travel that far north on the east side of the continent, while Rufous Hummingbirds travel right up to Alaska in the west, with Calliope and two others also travelling well into western Canada. Seven or so further species travel into the southern and central United States. The story of their migratory capacities is a legend in its own right, with Ruby-throated literally doubling its own lean body-weight of about 3.5 grams to put on enough fat to power a non-stop 800-kilometre flight across the Gulf of Mexico.

There are seventeen species of hummingbird in Trinidad, of which we saw ten. Curiously, only two of the seventeen bear the family name. Ruby-topaz Hummingbird, *Chrysolampis mosquitus*, has a scintillating scarlet head and gold throat to justify the jewel-like name, and I recall their streaks of coloured light glittering through the shrubs in the afternoon sun. The scientific name is a wonderful reflection of both the bird's luminescent plumage and its tiny size, since the generic refers to a 'glow-worm' and the specific to a 'little fly'.

Copper-rumped Hummingbird, *Amazilia tobaci*, tells of a somewhat less flamboyant garb with a metallic sheen. Reflecting such splendour, the generic name refers to an Inca princess in a 1777 novel by Jean Marmontel. The specific name reminds us that this species was first known by Latham as the Tobago Hummingbird – and, yes, there is a linguistic link between Tobago and tobacco.

The hummingbirds have beautiful, yet descriptive vernacular names and lots of fun in their scientific forms. With so many species to describe and name, the ingenuity and imagination of successive ornithologists have created some of the most intriguing names to describe their subtle variety. One of the tricks has been to use alternative words to replace 'hummingbird': as a consequence, Trinidad has representatives of the 'sabrewings', 'sapphires', 'emeralds', 'woodstars', 'violetears', 'starthroats' and 'goldenthroats', all of which indicate the capacity of this family to inspire – and these are but a tiny proportion of all the names created. Contrast such names with generally plain names of the wonderfully attractive Australian honeyeaters or the North American wood-warblers, and we realise what might have been.

The Tufted Coquette, *Lophornis ornatus*, takes the first part of its vernacular name from an outrageous orange punk hairdo and the second from the French for a 'flirty girl'. *Lophornis* describes it as 'a crested bird', while *ornata* reminds us that the male has the most magnificent ornate collar plumes. It too is a tiny, very active, insect-like species, and to see one well requires a lot of patience, though it is best seen at first light as it perches in the sun to warm up.

The word 'jacobin', as in the name of the White-necked Jacobin, *Florisuga mellivora,* is a little more of a mystery. It is a name most frequently used in the context of the political party of the 1789 French Revolution, but in this context it would seem to refer to the Dominican friars who were also known by the name 'Jacobins'. The bird's dark blue-green plumage, with white collar and white belly, has a passing resemblance to the dark cloak worn over a white habit, with the white hood exposed above the collar. Its scientific name reminds us that it is a 'honey-eating flower-sucker'.

It appears that garb also plays its part in a more common hummingbird name. The use of the word 'hermit' to describe some of the hummingbirds is a real paradox, especially when Richard ffrench writes that 'Hummingbirds are generally solitary by nature – apart from hermits,' Given that hermits in the human context are by definition solitary, this meaning is clearly not the reason for the use of that name. Rather it appears that hummingbirds bearing that name lack the lustrous plumage which characterises other species. In short, it is their relative drabness of garb which earns them the name. We saw Green Hermit, *Phaethornis guy,* which is literally 'Guy's sunbird', named after J. Guy, a nineteenth-century French naturalist. A second species was Rufous-breasted Hermit, *Glaucis hirsutus.* The generic name here refers to a 'pale greenish colour', but an alternative common name, Hairy Hermit, reflects the idea of the specific that the facial markings of the bird were likened to 'whiskers'.

Of all the pleasure we had in watching hummingbirds, no experience of them measured up to the morning when we were sitting immobile and silent under the forest canopy trying to spot a rarity. Bats were flying in the shadows and a few large insects were active. Suddenly a loud buzz echoed round my head. Fearing a large hornet or the like I flinched and was halted from swiping instinctively at my face by the urgent whisper from someone behind me. Just as well. It was in fact a Rufous-breasted Hermit, which must have confused my left ear with some exotic bloom. I recalled a Barn Owl and my father's words: 'Sit still and you just don't know what will happen.' That was a magic moment, but my father was no longer around to hear the tale.

We saw two species of mango, Black-throated and Green-throated. These belong to the same genus as *Anthracothorax mango,* which was originally referred to as the Jamaican Mango Hummingbird. Mango trees were introduced into Jamaica in 1782, after the British Navy had captured a French ship which was carrying some plants and seeds, and that was the start of an important local enterprise. It seems that the hummingbird later became associated with the fruit, that the name stuck, and that it was then attached to related species. Black-throated is *A. nigricollis,* literally 'black-collared', while Green-throated, *A. viridigula,* is a more exact match.

Not-so-plain names

As in Australia, there are plenty of vernacular names, which appear slightly unusual because, to the British birder, they create unfamiliar associations of familiar words, but they are nonetheless easily comprehensible. Words like woodcreeper, leaftosser, greenlet, spinetail, seed-eater, and peppershrike all offer a touch of the exotic in that way, but they tend to be simple descriptive names which tell something useful. The Bicolored Conebill seems a little more extravagant, but it is still easily interpreted.

What is striking among the names created in this mode is the importance of one insect in supporting a whole range of birds. The humble ant is apparently the mainstay of antshrike, antvireo, antwren, antthrush, antbird, antpitta and ant-tanager, and each of these may include a number of species. The secretive Scaled Antpitta led us a merry chase, even though it was calling in nearby undergrowth, but we saw some other species well, such as the Barred Antshrikes which were nesting by our door.

PIED WATER TYRANT

I discussed the tyrant-flycatchers briefly in my section on North American bird names, but we saw a number of them in Trinidad, thanks largely to the skill of our guides – some eighteen species in all. There were some easy ones, like the Tropical Kingbird, Great Kiskadee and Boat-billed Flycatcher mentioned earlier. Pied Water Tyrant, *Fluvicola pica*, lived up to its common and scientific names in being a black and white bird found near water. White-headed Marsh Tyrant, *Arundinicola leucocephala*, lived up to its name in a similar way, the 'marsh' part of the vernacular name being conveyed by the generic name, meaning 'reed-dweller'.

The Southern Beardless Tyrannulet, *Camptostoma obsoletum*, provided a new element, with the name 'tyrannulet' befitting a small member of the family. *Camptostoma* simply means 'curved bill', but 'beardless' is explained in the name of the Northern equivalent, *C. imberbe*, which refers to the lack of rictal bristles. The Southern species is *C. obsoletum*, which in this case may presumably be taken literally as 'worn out' with reference to the same non-existent bristles.

In general this group contained many little confusing greyish, olive or brownish species, like the Dusky-capped Flycatcher, but two species carried 'greenness' in their generic name *Elaenia*, which relates to the Greek for 'oily', in this case 'olive green'. Not that it helped too much when the choices included an Olive-sided Flycatcher on its passage northwards, or where the forest shadows made them all so indistinguishable, but expert help was at hand. In retrospect I value more highly the single Yellow-bellied Flycatcher that I found for myself in Cape Breton, Nova Scotia.

Two relatives of the Brown Cowbird appeared during our visit. The first was Shiny Cowbird, *Molothrus bonariensis*, which we saw at the Aripo cattle station, in open country. This bird is clearly very widespread, since it takes its specific name from Buenos Aires in Brazil. The second species was the Giant Cowbird, *Molothrus oryzivorus*, which was initially spotted by those travelling in the vehicle ahead. The intercom crackled to alert us. Of all such shorthand messages, 'Giant Cowbird on Water Buffalo' has to be one of the most surreal bits of birding news I have ever heard. The episode would not have struck me as quite so bizarre had I previously read Richard ffrench's comment on the species' food: 'Invertebrates taken on the ground or off the hides of domestic animals.' By the time we reached the spot, Giant Cowbird was *under* Water Buffalo, illustrating the first of those feeding habits. The bird has a somewhat monumental name to match its size. I dealt with *Molothrus* earlier, when discussing the the Brown-headed Cowbird. The specific name *oryzivorus* means 'rice-eating', which explanation allows me to complete the quotation of ffrench's note: 'also grain, including rice.'

***

In conclusion, it is quite clear that the names of the birds of Trinidad represent a much wider range of languages and cultures than those of either North America or Australia. It is refreshing to see that so many elements reflect and respect the names used by the Amerindian peoples who inhabited the region before colonisation, even if those roots are more widely South American than Trinidadian. The languages of Portugal, Spain and France are widely represented, as is Dutch, which is no great surprise. It has to be borne in mind, however, that many of the names discussed are still those agreed by an international English-speaking community of ornithologists, and in that respect they tend to override or ignore many of the forms used by the people of Trinidad. I acknowledge such local names, but leave most to one side, since they are outside the scope of my intentions. More detail of them will be found in Richard ffrench's or in Lise Winer's work.

JOURNEY'S END

And so I come towards the end of my journey through what once appeared to be a pathless wilderness.

When I first started to explore the confusions and contradictions found in the naming of birds, I felt that maybe it was a question of 'them' not getting their act together, but of course it is not as simple as that. 'They' are collectively the players in a gigantic game, which involves about 10,000 species of birds; umpteen different nationalities, cultures and subcultures; the vagaries of usage and linguistic evolution; the ongoing discoveries of scientists; the micro- and macro-politics of institutions; the decisions of authors and editors; the passage of time; the fashions of an age; the passions and opinions of individuals... In short, the causes and effects which shape the choice of bird names are just about as complicated as any other area of human activity. Those who anticipate a definitive, fixed and authoritative solution to the universal naming of all birds will, I am sure, be waiting more in hope than in expectation, but we do in fact already seem closer to a consensus on international English names than when I started to sort out my Black-bellied Plovers and Grey Plovers in Canada in 1994, which moment was my first step on this particular trail.

For me the story of the names underlines the timelessness of human interest in the natural world. We still have the same curiosity as Aristotle; we still need to understand, to admire, to appreciate and even to label. In understanding and labelling we fulfil a rational need, but in appreciating and admiring we are satisfying the needs of our emotional and spiritual being. But what is wonderful is that over twenty-three centuries after the observations of Aristotle, we still don't know all the answers, because the natural world is so complex.

Sadly the twenty-first century will provide massive challenges in terms of climate change, habitat loss, pollution, resource consumption, and many other issues. With mass extinctions looming, the big question will be how long the natural world can survive our species' growing need for food and our lust for material benefits. In short, will there still be birds to name? But for now they are there and the process of understanding them goes on.

As in the world of birding itself, there are more linguistic trails to follow in other parts of the world, but for now, at least, I need to sort and pack away the paraphernalia of this journey. It has been immense fun, sometimes tiring, and often very eye-opening. I just hope that you were not lost along the trail.

APPENDIX
THE LEGENDS BEHIND THE NAMES

Is this the legend of Gerana, 'an eternal feud between the Cranes and the Pygmies'?

MOSAIC DETAIL FROM ITALICA

During the writing of this book, I was often conscious that the requirements of the narrative allowed me only to make a general reference to the Classical legends and stories that occur so frequently in the scientific nomenclature. It is entirely probable that such references would have been much less frequent if the exercise of naming of birds had begun in the twenty-first century, because the names are largely the product of a time and culture in which Latin and Greek were at the core of European education. Nonetheless, the background stories are fascinating in themselves. This appendix therefore aims to provide a little more depth to some of the links between the legends and the birds.

AVIAN TRANSMUTATIONS IN THE CLASSICAL WORLD

Numerous legends of Greece and Rome lead to the conversion of aggressors, victims and bystanders into birds. The profound influence of animistic religious beliefs is at the root of this tradition, and the concept has its parallels in many other cultures, where people or their spirits become birds after death. The wings of angels are linked to this tradition. In short, transmutations are a symbolic interpretation of deaths which result from a tragedy, and a device to promote a belief in life after death.

The dynasty of Pandion I (Osprey, Swallow, Nightingale, Song Thrush, Sparrowhawk and Lapwing)

The relationships in this family of legends are complex. What is certain is that the dynasty was more than susceptible to the whims of the gods, since most seemed to finish up as some sort of bird.

Pandion I had three children: Pandion II, who fathered Nisus, whose daughter was Scylla; Procne/Progne, who married Tereus; and Philomela, who was raped by Tereus.

The story of Nisus and Scylla

The rule of King Nisus, son of Pandion II, was ended when King Minos of Crete invaded Nisus's kingdom. Nisus himself was protected by a lock of purple hair which guaranteed his invincibility, but was betrayed by his daughter Scylla, who became besotted with Minos (or was bribed in some versions). When Scylla cut off the magic curl, Nisus was defeated and killed, but was transformed into a bird of prey – an Osprey/sea eagle or hawk, according to various versions, and possibly dependent on the interpretation of translators. He then drowned his errant daughter as she tried to scramble aboard the Cretan king's ship. She in turn transformed into a lark/seabird/heron (the versions again vary) destined to be pursued forever by the bird of prey.

The story of Tereus, Procne and Philomela

The legend of Tereus was told in drama by the Greeks, Sophocles and Philocles, and by the Roman, Ovid, in Book 6 of his *Metamorphoses*.

The essence of the story is similar each time, but an important difference lies in the transmutation of Tereus.

Tereus lusts after his sister-in-law Philomela and, after ravishing her, cuts out her tongue to silence her and leaves her imprisoned in the woods. She literally embroiders a version of the story, which she sends to her sister, Procne. In vengeance for her husband's perfidy, Procne kills their son and serves the roast infant for her husband's dinner. In fury, Tereus pursues both women and, during the pursuit, the gods transmute the trio into birds. Procne becomes a Swallow, Philomela a Nightingale, and Tereus a hawk, according to the Greeks, but a Hoopoe according to Ovid.

Given the complications of differing versions of these legends, and the fact that Pandion II may have been invented later to fill in a perceived gap in the legendary monarchy, it is not surprising that some confusions occurred. Savigny's error in allocating Pandion's name to the Osprey simply seems to have been a tale half-remembered from a drowsy Friday afternoon lesson. Linnaeus himself appears to make a similar error with Antigone and the Sarus Crane (see later).

The fact that Nisus eventually becomes the Sparrowhawk also seems a little odd, given the legend's maritime context, where such as an Osprey might have been expected. However, it must be remembered that such legends should be seen as 'embroidered' history, and that detail was probably distorted in transmission and translation.

Ovid's decision to introduce a Hoopoe in preference to a bird of prey was no error. To modern eyes the Hoopoe seems to be a feeble symbol of aggression, inappropriate to the vengeful spirit of the original Greek version. In fact, the idea is rooted in Aristophanes' much older comedy, *The Birds*, where the Hoopoe is the king, the crest representing the crown, and the long, strong bill symbolises a violent nature. Ovid's version then leads to the coincidental slandering of the Lapwing as 'treacherous', simply because it too sports a crest.

Only Procne and Philomela seem relatively straightforward in their attachment to the Swallow and the Nightingale, while Scylla has no more to do with the modern naming of birds. In modern nomenclature the traces of Procne/Progne are found in the several scientific names of swallows and martins, but also in names of terns and some other species with swallow-like characteristics. The genus *Philomela* has been superseded by *Luscinia*, but a form still exists as the specific name of the Song Thrush, *Turdus philomelos*.

The servant girl earning eighteen pieces (Collared Dove)

According to James Fisher, in an article in *British Birds* in May 1953, the legend was told by C. Hinke to J. F. Naumann in 1837, the year before Frivaldszky named the bird. His version goes thus:

> *A poor maid was servant to a hard-hearted mistress, who gave her as wages no more than eighteen pieces a year. The maid prayed to the Gods that she would like it to be known to the world how miserably she was paid by her mistress. Thereupon, Zeus created this Dove, which proclaims an audible deca-octo to all the world to this very day.*

The flaw in that story is that *decaocto* is a four-syllable structure, while the dove has a three-note song. Perhaps *dec'octo* works. Fisher's article also gives an account of the dramatic spread of the bird.

Arne of Thrace (Jackdaw)

The story of Arne Sithonis (meaning that she originated from Thrace) appears in Ovid's *Metamorphoses*, written in about 8 BCE. Ovid uses second-hand material gleaned from a culture much older than his own and tells the story in his own way.

This is the tale of the betrayal of Arne's adopted island nation by her love for money. She accepts bribes from King Minos of Crete, who then attacks her island. Her punishment by the gods is her transmutation into an acquisitive Jackdaw, as a symbol of greed, hence the specific name *monedula* (money-eating).

The bare bones of this story have more than a touch of déjà vu, since it was the same King Minos who seduced Scylla and brought down the wrath of the conquered King Nisus. The coincidence can probably be explained by the fact that legends are often folk morality tales transmitted by a largely oral tradition and attached to convenient historical moments to give them some authenticity. The Celtic legends of King Arthur are in much the same mode. The coincidence of the betrayal by women of two nations to the same King Minos suggests very strongly that both stories have a common root. It seems that Classical historians have long discussed exactly which island was involved in the Arne legend.

Alcyone and Ceyx (kingfishers)

In Greek legend, Alcyone, daughter of Aeolus, god of the wind, and Ceyx, son of Eosphorus, the Morning Star, lived in wedded bliss, so happy that they called each other by the pet names of Hera and Zeus. This blasphemy offended the god Zeus so much that he sank the ship taking Ceyx to consult the Oracle. When the corpse of her husband washed to the shore, Alcyone killed herself in a fit of despair, at which point the gods were distraught, Zeus was contrite, and the couple were turned into kingfishers.

Atthis (Kingfisher)

Ovid tells the story of Atthis in Book V of the *Metamorphoses*. In it the resplendent Indian prince, son of the Ganges water-nymph Limniace, becomes embroiled in a quarrel, during which Perseus (better known as the slayer of the Gorgon and founder of Mycenae) kills him.

The family of Autonus: Anthus (Pipit), Acanthis (Redpoll), Erodias (Heron) and Schoenus (Reed Bunting and Sedge Warbler)

Autonus and his wife Hippodamia had five children: Anthus, Erodius, Schoenus, Acanthis and Acanthus. The family owned poor land and a large herd of horses. Erodius tended his father's horses on the best pasture, but one day Anthus drove them into a poorer area. The hungry horses attacked Anthus: in one version they ran him down and, less credibly, in another version they devoured him. Pitying the stricken family, the gods transformed them all into birds.

Several of the family remain important in the names in current use:

- Anthus himself became a bird which neighed like a horse, but always fled from horses. From this comes the genus name for pipits.
- Acanthis became a finch, and thus the name was chosen for the redpolls.
- Erodius became a heron, and reappears as *Ardea herodias*, the specific name of the Great Blue Heron.
- Schoenus (meaning 'reed') appears in Aristotle as *skhoiniklos*, a 'small waterside bird' and reappears today in such specifics as *schoeniclus* (Reed Bunting) and *schoenobaenus* (Sedge Warbler). Strictly, both Schoenus and the birds take their name from the plant.

Diomedes (albatrosses)

Diomedes was king of Aetolia, a major figure in Homer's version of the Trojan Wars. He contributed the third largest fleet to the Greek expedition against Troy. Later he found that he had been betrayed by his wife in his absence, so abandoned his kingdom to found the city of Argyrippa in Italy, where he eventually died. In brief, he was reputedly a great warrior, a cunning leader and a pragmatist who drew the unstinting loyalty and devotion of his followers. After his death he became a demigod.

The story of his death, which pertains to the naming of the albatross genus *Diomedea*, contains contradictions. Homer's version has albatrosses gathering to sing mournfully at the news of his death, while Roman versions have his followers transmuting into seabirds.

Jobling points out that the *aves Diomaedias* of the Roman authors were probably gulls or gannets and then, somewhat enigmatically, brackets the word *Calonectris* at the end of his note. For those readers who may have bristled at the anachronistic word 'albatrosses' used to recount the Homerian version, it has to be realised that this would be a usage of much more recent translators, an 'approximate' word, and one which was imbued with much mystique by more modern poets, such as Coleridge and Baudelaire. The *Calonectris* species are the 'noble shearwaters'. What we now know as Scopoli's Shearwater, a relative of the albatrosses, might well have been at the wake.

Neophron (Egyptian Vulture) and Aegypius (Cinereous Vulture)

This legend was recorded in the *Metamorphoses* of the Greek, Antoninus Liberalis (second or third century CE). He recycled a much older legend which tells a complicated story of filial perfidy. Neophron is upset by the relationship between his widowed mother, Timandra, and the much younger Aegypius. He betrays them by luring the rival's mother, Bulis, to his house in the absence of Timandra and placing her in a darkened room just before the arrival of Aegypius. When the mother becomes aware that she has made love to her own son, she blinds him and kills herself. The gods turn the two men into vultures and Bulis into a bird which pecks the eyes from fish (some have it as a heron). Timandra becomes a small bird, possibly a Long-tailed Tit.

The Muses (Magpie) and Calliope (Siberian Rubythroat)

The Greek Muses personified the knowledge and skills of the arts. Traditionally there were nine Muses, daughters of Zeus and Mnemosyne (though some traditions had only three). Some versions tell that they were also known as the Pierides, from the region of Pieria, whence they came. That fact did not help the case of the daughters of King Pieris (see below).

In Roman times the nine were attributed to different disciplines, Calliope being attached to epic poetry. Clio represented history; Euterpe, flutes and lyric poetry; Thalia, comedy and pastoral poetry; Melpomene, tragedy; Terpsichore, dance; Erato, love poetry; Polyhymnia, sacred poetry; and Urania, astronomy. As the chief of the Muses, Calliope was reputed to have the most beautiful voice, and she became the mother of Orpheus.

Ovid recounts a clash which ended in a spectacular transmutation. King Pieris of Macedon had nine daughters who, because of his name, were also known as the Pierides. Foolishly he challenged the Muses to a contest. His daughters proved to be unworthy rivals, so his brazen nerve was punished when his daughters were changed into chattering Magpies.

The original English and French form of Pie for Magpie derives from the Latin pica, which is based on the bird's clucking call, so any apparent linguistic link to the Pierides is merely coincidental.

The story of Picus (woodpeckers)

In Roman mythology, Picus was the first king of Latium. His first mistake appears to have been that he made the wood-nymph, Canens, his wife, which action upset the witch Circe. His second mistake was to spurn the advances of the witch herself. The result was that she turned him into a woodpecker, at which Canens killed herself.

Antigone of Thebes, Antigone of Troy, and Gerana, Queen of the Pygmies (Sarus Crane)

The story of Antigone of Thebes was dramatised by Sophocles in the fifth century BCE. It was resurrected during the Nazi occupation of France in World War 2 by Jean Anouilh, as a vehicle for the discussion of authority, moral values, tyranny and defiance. As originally told by Sophocles, Antigone was the daughter of Oedipus and Jocasta, born from the tragic union of son and mother. Her two brothers fought to the death over the succession to the throne, which was then in the hands of their uncle Creon. The loyal brother was buried with honour and the perceived traitor was left to the scavengers on the battlefield. Antigone defied her uncle in attempting to bury her disgraced brother, and the confrontation led (in Sophocles) to her suicide by hanging.

The more credible story links the Sarus Crane to Antigone of Troy, a much less well-known figure. She was a daughter of Laomedon and sister to Priam. Her story is a parable of vanity, since she claimed to have hair more beautiful than that of the goddess Hera. In a move reminiscent of the vengeance of Zeus on Ceyx and Alcyone, Hera changes the mortal's hair into snakes, but a merciful god changes her instead into a stork – which of course preys on snakes.

However, in a parallel myth, Gerana, Queen of the Pygmies, also claims to be more beautiful than Hera, but this time the goddess turns her straight into a crane, a move which provokes an eternal feud between the Cranes and the Pygmies. Gerana's name, by the way, is based on *géranos*, the Greek for Crane (though she is sometimes known as Oenoe).

The linking of the first story with the Sarus Crane is based only upon the red neck of the bird: it is tenuous in the extreme. It seems a touch too facile to probe no deeper than the surface of the much better-known story. On the other hand, the stories of Antigone of Troy and of Gerana have so much in common that it is not so hard to imagine Linnaeus dredging his memory and confusing two very similar stories before allocating the wrong bird to the wrong woman. It is, after all, not a unique example of his fallibilities and is exactly the same sort of error as that made by Savigny in naming the Osprey *Pandion*.

OTHER LEGENDS

Some stories cited in the text do not involve transmutation. The legends of this second group often reflect physical or behavioural characteristics.

Swan song

The origins of the legend of 'swan song', a beautiful final melody from the dying bird, are rooted in their dedication to Apollo as a symbol of harmony and beauty. Curiously the myth was perpetuated by Aristotle, though refuted firmly by Pliny the Elder.

Penelope (Wigeon)

Legend has it that Penelope was the daughter of Prince Icarius of Sparta and the nymph Periboea. At her birth her mother hid the child from her father, who, typical of the warrior class which dominated the society, had wanted a son. When Icarius discovered the child, he threw her into the waves, but she was rescued by ducks, spared by her father, and given her name after the ducks (Greek *penelops*). It was Linnaeus who associated the Eurasian Wigeon with that specific legend.

Later Penelope was to become the wife of Odysseus and to become a legend for her fidelity and for the cunning by which she stayed the advances of many suitors during the twenty-year absence of her husband. An alternative source of the meaning of her name derives from the weaving with which she is associated, but for the purposes of the bird's name, the duck story is clearly preferable.

Gryphon (vultures)

Alternative spellings (*griffin, griffon* and other archaic variants) are used for this mythical creature, which is associated with legends of gold in Scythia (modern Khazakhstan). One theory suggests that fossil remains were at the root of a legend which was widespread in Greece, Persia and Egypt and was frequently represented in art and literature. The beast was considered capable of extracting gold from the rocks and was guardian of hoards of gold. Half-eagle, half-lion, and gigantic, it reigned supreme: according to Philostratus, it was capable of preying on even the elephants. In post-classical literature it appears in works by such as Dante, Milton and Lewis Carroll and has served frequently as a heraldic device (though wingless in English heraldry).

Cecrops (Red-rumped Swallow)

Cecrops was the mythical founder of Athens, often depicted as half-man, half-serpent. The early Athenians believed that he had imparted the norms of civilised behaviour, which were the foundations of their later success. He is also associated with the Acropolis, having judged Athena to be the winner of a race with Poseidon to possess the rock, which was also known as the Cecropia in his honour.

Athena (Little Owl)

Many legends tell differing versions of the origins of the goddess Athena, but the Olympian story is that she was born from the head of Zeus himself. Like Cecrops, Athena was in many ways an overarching representation of all that amounted to civilised behaviour – the epitome of wisdom, civilisation, virtue and so on. Such values led to the success and enduring fame of an Athenian culture which invented the concept of democracy and remains an example to the modern world. Athena's temple, the Parthenon, was erected on the highest point in the city of Athens, which took its very name from her. From the mundane story of the Little

Owl nesting in the temple derived the use of the bird as a symbol for the cults of Athena and of Minerva, her 'successor' in Ancient Rome.

Orpheus (Orphean Warbler)

In Ancient Greek myth, Orpheus, son of Calliope, the chief Muse, was a renowned poet and musician, who was said to have perfected the lyre. His music literally charmed all around him, including the birds and beasts of the field. One poet suggested that he could move even trees and rocks. A cult arose from his reputation as a prophet. From this sprouted a number of oracles.

The three Graces – Aglaïa (Rose-throated Becard)

The three Graces, or Charities, were generally considered to be the daughters of Zeus and Eurynome, though other versions propose alternative divine parentage. Aglaïa, the youngest, represented Splendour. Eurosyne was Mirth, and Thalia was Good Cheer. To the Romans they were the *Gratiae*.

Jason and the Argonauts (Pheasant)

The story of Jason's adventures is possibly the best known of the legends of Ancient Greece, having been the subject of film and television versions. It was first recounted by Apollonius Rhodius in the third century BCE as the *Argonautica*. In short, Jason attempts to reclaim the throne from his uncle and, to prove himself, is challenged to sail to Colchis and bring back the Golden Fleece. A huge variety of adventures occurs before Jason succeeds, aided by the sorceress Medea, whom he marries. Somehow, in between mastering fiery oxen, defeating an army grown from a dragon's teeth, and evading the dragon itself, the Argonauts apparently remembered to ship a few of the local pheasants and thus to introduce them to the western world. Undoubtedly, the introduction was much more mundane in reality, but the links with the land Colchis and the River Phasis are there in both the legend and the names of the bird.

The Harpies (Double-toothed Kite and Harpy Eagle)

In Greek legend, the Harpies were winged female demons whose role was to torment Phineus, who had offended Zeus with his gift of prophesy. Blinded, Phineus had to endure the repeated attacks of the Harpies, who snatched his food. His ordeal was relieved only when the Boreads, sons of the North Wind and members of Jason's Argonauts, drove the Harpies away. In return, the crew was aided by Phineus in their passage through the Clashing Rocks, the Symplegades. The image of the Harpies had widespread currency in Greece and Rome, and lingered into more modern times to appear in such as Dante's *Inferno*. In modern English 'harpy' is still a derogatory term for a spiteful woman

Birds of augury: Perisoreus (Grey Jay and Siberian Jay)

The Augurs of Ancient Rome were soothsayers, who linked the signs of the natural world to events and occurrences. Some of their principal interpretations were *ex avibus*, i.e. based on the behaviour of certain birds. Such interpretations depended on the species concerned, and in particular on either its voice or its direction of flight. According to Elliott Coues, the *–soreus* element of *Perisoreus* was based on *sorix*, a bird of augury dedicated to Saturn. His sole motive for this claim appears to be that the Siberian Jay was considered by some in Europe to be a bird of ill-omen. The idea that *Perisoreus* was to do with the bird's hoarding habits seems rather less contrived and more credible.

The Cyclops (Double-eyed Fig Parrot)

The Cyclopes were a mythical race of one-eyed giants, often depicted with an eye in the centre of the forehead. In Homer's *Odyssey*, Odysseus arrives in a remote land, where he meets and slays the Cyclops Polyphemus, a son of Poseideon. Homer's version becomes the pattern for later interpretations.

The Satyrs (Tityra)

In Greek legend and drama the Tityri were Satyrs, half-man, half-goat, and followers of Dionysus, the god of wine. They symbolised debauched behaviour and excess. The name Tityri comes from the double flutes (*tityrinos*) with which they are normally depicted, but the South American birds have a raucous voice, so earn their name by their unruly and aggressive behaviour.

The Oceanides (Wilson's Storm Petrel)

The Oceanides were said to be the daughters of the Titans, Oceanus and Tethys, and were 3,000 in number, each one a patroness of some natural feature, particularly of sources of water. In spite of the appearance of the name to modern eyes, both Oceanus and his offspring were originally associated with fresh water, but later the encircling river represented by Oceanus became a metaphor for the Atlantic Ocean, a fact which is more appropriate to the name of a storm petrel species.

The Nereids (Fairy Tern)

The Nereids were sea-nymphs, the fifty daughters of Nereus and Doris, and particularly associated with the Aegean Sea, where they lived in a silver cave. They accompanied Poseidon and gave succour to storm-stricken sailors. Some versions have the Nereid, Amphitrite, as the wife of Poseidon and another of their number was Thetis, mother of Achilles.

Phaetusa (Large-billed Tern)

Phaetusa was, according to some, the daughter of the Greek god Helios and supposedly in charge of his flocks. The exact status of Helios, a secondary god, was sometimes confused with that of Phoebus/Apollo, who was chief among the gods alongside Zeus. Phoebus is often linked with the sun, as in Chaucer's lines to describe dawn:

> *The bisy lark, messager of day*
> *Salueth in hir song the morwe gray*
> *And firy Phoebus riseth up so bright*
> *That al the Orient laugheth of the light.*

BIBLIOGRAPHY

Attenborough, D. (1998) *The Life of Birds*. London: BBC Books.

Avery, M. (2014) *A Message from Martha: the Extinction of the Passenger Pigeon and its Relevance Today*. London: Bloomsbury.

Bircham, P. (2007) *A History of Ornithology*. New Naturalist 104. London: Collins.

Birds of the Western Palearctic Interactive. BirdGuides Ltd 2004–2006 (BWPi). http://www.birdguides.com/bwpi.

Balmer, D. E., Gillings, S., Caffrey, B. J., Swann, R. L., Downie I. S. and Fuller, R. J. (2013) *Bird Atlas 2007–11: the Breeding and Wintering Birds of Britain and Ireland*. Thetford: BTO Books.

Beolens, B. and Watkins M. (2003) *Whose Bird? Men and Women Commemorated in the Common Names of Birds*. London: Christopher Helm.

Bonta, M. (2010) Transmutation of human knowledge about birds in 16th-century Honduras. In Tidemann S. & Gosler A. (eds.) *Ethno-Ornithology: Birds, Indigenous Peoples, Culture and Society*. Oxford: Earthscan, pp. 89–102.

Bosworth Smith, R. (1913) *Bird Life and Bird Lore*. Cambridge: Cambridge University Press.

British Ornithologists' Union (2013) The British List: a checklist of birds of Britain (8th edition). *Ibis* 155: 635–676.

Brown, L. (1982) *British Birds of Prey: a Study of Britain's 24 Diurnal Raptors*. London: Collins.

Bucknell, N., Clews, B., Righelato, R. and Robinson, C. (2013) *The Birds of Berkshire: Atlas and Avifauna*. Reading: BBAG.

Burnham, K. and Newton, I. (2011) Seasonal movements of Gyrfalcons. *Ibis* 153: 468–484.

Campbell, B. (1974) *The Dictionary of Birds in Colour*. London: BCA/Michael Joseph.

Centre National des Ressources Textuelles et Lexicales (on-line etymological French dictionary). http://www.cnrtl.fr/etymologie/

Choate, E.A. (1973) *The Dictionary of American Bird Names*. Boston, MA: Gambit.

Cocker, M. & Mabey, R. (2005) *Birds Britannica*. London: Chatto and Windus.

Collar N. (2013) A species is whatever I say it is. *British Birds* 106: 130–142.

Coward, T. A. (1920) *The Birds of the British Isles and Their Eggs* (First and Second Series). London: Frederick Warne.

Coward T. A. (1926) *The Birds of the British Isles: Migration and Habits* (Third Series). London: Frederick Warne.

Cumming, W. P., Hillier, S. E., Quinn, D. B. and Williams, G. (1974) *The Exploration of North America 1630–1776*. London: Paul Elek.

Darwin, C. (1968) *The Voyage of the Beagle*. Geneva: Heron Books.

Darwin, C. (1979) *The Origin of Species* (abridged R. E. Leakey). London: BCA / Faber and Faber.

Dictionnaire Littré en Ligne, Reverso. littre.reverso.net/dictionnaire-francais.

Elphick, C., Dunning, J. B., Sibley D. (2001) *The Sibley Guide to Bird Life and Behaviour*. London: Christopher Helm.

Ferguson-Lees, J. & Christie, D. A. (2001) *Raptors of the World*. London: Christopher Helm.

ffrench, R. (2003) *A Guide to the Birds of Trinidad and Tobago*. London: Christopher Helm.

Fisher, J. (1966) *The Shell Bird Book*. London: Ebury Press and Michael Joseph.

Fisher, J. (1953) The Collared Turtle Dove in Europe. *British Birds*, 46: 153–181.

Game and Wildlife Trust. http://www.gwct.org.uk/game/

Génsbøl, B. (1987) *Collins Guide to the Birds of Prey of Britain, Europe, North Africa and the Middle East*. London: Collins.

Gill, F. and Donsker, D. (eds) (2016) *IOC World Bird List* (v 6.1). http://www.worldbirdnames.org/ioc-lists/master-list-2.

Goodfellow, D. L. (2005) *Birds of Australia's Top End*. Sydney: Reed New Holland.

Hartert, E., Jourdain, F. C. R., Ticehurst, N. F. and Witherby, H. F. (1912) *A Hand-List of British Birds*. London: Witherby and Co.

Hilty, S. L. (2003) *Birds of Venezuela*. London: Christoper Helm.

Holloway, S. (1996) *The Historical Atlas of Breeding Birds in Britain and Ireland: 1875–1900*. London: Poyser.

Horne, A. (2009) *Birdwatching Watching*. London: Virgin Books.

Jobling, J. A. (2010) *Helm Dictionary of Scientific Bird Names from Aalge to Zusii*. London: Christoper Helm.

Juniper, T. (2002) *Spix's Macaw*. London: Fourth Estate.

Kearton R. (1900) *Our Bird Friends*. London: Cassell.

Linnaeus, C. (1758) *Systema Naturae per regna tria naturae*, 10th edition, openlibrary.org, Internet Archive. http://archive.org/stream/carolilinnsy01linn.

Lockwood, W. B. (1984) *The Oxford Book of Bird Names*. Oxford: Oxford University Press.

Löfgren, L. (1984) *Ocean Birds: their Breeding Biology and Behaviour*. London and Canberra: Croom Helm.

Macdonald, J. D. (1987) *The Illustrated Dictionary of Australian Birds by Common Name*. Frenchs Forest, NSW: Reed Books.

Madge, S. and McGowan P. (2002) *Pheasants, Partridges and Grouse*. London: Christopher Helm.

McCarthy, M. (2009) *Say Goodbye to the Cuckoo*. London: John Murray.

Morris, F. O. (1850) *British Birds* (ed. T. Soper, 1985). London: Peerage Books.

Naish, D. (2010) The incredible bill of the Oystercatcher. *ScienceBlogs*. scienceblogs.com/tetrapodzoology/2010/07/11/incredible-bill-of-oystercatcher/

Perrins, C. (ed.) (2003) *The New Encyclopaedia of Birds*. Oxford: Oxford University Press.

Pizzey, G. and Knight, F. (2007) *The Field Guide to the Birds of Australia*, 8th edn. Sydney: HarperCollins.

Reedman, R. (2013) The origins of the vernacular name of the Common Scoter. *British Birds* 106: 417.

Robinson F. N. (ed.) (1957) *The Works of Geoffrey Chaucer*, 2nd edn. Cambridge, MA: Riverside Press.

Sangster, G., Collinson, M., Crochet, P.-A., Knox, A. G., Parkin, D. T. and Votier, S. C. (2012) Taxonomic recommendations for British birds: eighth report. *Ibis* 154: 874–883.

Scott, P. M. (1980) *Observations of Wildlife*. Oxford: Phaidon.

Sibley, D. A. (2014) *The Sibley Guide to Birds*, 2nd edn. New York, NY: Alfred A. Kopf.

Simpson, K. and Day, N. (1998) *Field Guide to the Birds of Australia*. London: Christopher Helm.

Soper, T. (1985) *Tony Soper's Birdwatch*. Exeter: Webb and Bower.

Stoll N. R. (ed.) (1961) *The International Code of Zoological Nomenclature adopted by the XV International Congress of Zoology*. London: The International Trust for Zoological Nomenclature.

Svensson, L., Mullarney, K. and Zetterström, D. (2011) *The Collins Bird Guide*, 2nd edn (large format edition). London: HarperCollins.

Van Grouw, K. (2013) *The Unfeathered Bird*. Princeton, NJ: Princeton University Press.

Walters, M. (2003) *A Concise History of Ornithology*. London: Helm.

Warham, J. (1990) *The Petrels: their Ecology and Breeding Systems*. London: Academic Press.

Watson, D. (1977) *The Hen Harrier*. London: Poyser.

Wernham, C. V., Toms, M. P., Marchant, J. H., Clark, J. A., Siriwardena, G. M. & Baillie, S. R. (eds) (2002) *The Migration Atlas: Movements of the Birds of Britain and Ireland*. London: Poyser.

White, G. (1789, revised 1993) *The Natural History of Selborne*. London: Thames and Hudson.

White, T. H. (1963) *The Goshawk*. London: Penguin.

Winer, L. (2008) *Dictionary of the English/Creole of Trinidad & Tobago: On Historical Principles*. Montreal: McGill Queen's University Press.

Yalden, D. W. and Albarella, U. (2009) *The History of British Birds*. Oxford: Oxford University Press.

INDEX

Words in parentheses refer to those names which are archaic, purely local or obsolete